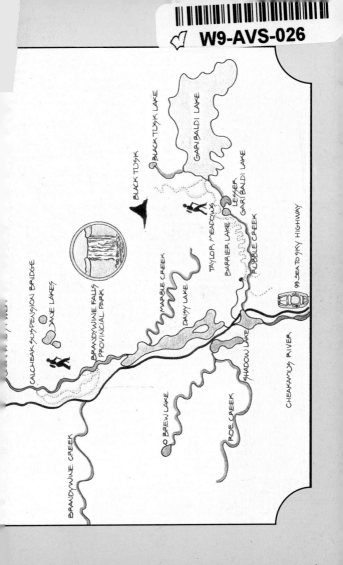

LARGE-LEAFED AVENS

Geum macrophyllum • Rose family: *Rosaceae*

■ **DESCRIPTION** Large-leafed avens is a herbaceous perennial to 90 cm in height. Its bright yellow flowers resemble butter-cups. They are approximately 6 mm across and are produced singularly or in small clusters. The unique round fruit has bristly bent protruding styles that catch on fur and clothing, an excellent way of dispersing the seed. The irregular-shaped larger leaves are 15-20 cm across, while the stem leaves are smaller and 3-lobed.

■ **HABITAT** Prefers moist soil in open forests and beside pathways, trails and roads at low elevations.

■ **NATIVE USE** The roots were boiled and used medicinally.

■ **LOCAL SITES** Lines the pathway around Alpha Lake; trailside to Rainbow Falls, Brandywine Falls and Whistler Village. Flowers May to August, depending on elevation.

39

FAN-LEAFED CINQUEFOIL

Potentilla flabellifolia • Rose family: *Rosaceae*

■ **DESCRIPTION** Fan-leafed cinquefoil is a herbaceous perennial to 30 cm in height. Its yellow flowers to 2 cm across have 5 petals that surround a central button. The leaves have 3 leaflets that are deeply notched and spread out like a fan. The genus name *Potentilla* means "powerful," a reference to its medicinal properties. Infusions of *Potentilla* were used in Europe for sore throats, cramps in the stomach, heart and abdomen. The species name *flabellifolia* means "fan-shaped."

■ **HABITAT** Moist slopes and meadows at high elevations.

■ **LOCAL SITES** Common in the subalpine slopes and meadows. Can be seen flowering July to August to Seventh Heaven, Blackcomb Mountain.

PARTRIDGEFOOT

Luetkea pectinata • Rose family: *Rosaceae*

■ **DESCRIPTION** Partridgefoot is a mat-forming miniature subshrub to 12 cm tall. Its cream-coloured flowers form in compact terminal clusters to 2 cm long. The evergreen leaves are finely dissected and resemble the feathered leggings on grouse and ptarmigan. The genus *Luetkea* honours Count Friedrich Luetke, a supporter of the St. Petersburg Academy of Sciences and commander of a Russian ship that charted the coast of Alaska in 1827.

■ **HABITAT** Meadows and slopes at subalpine to alpine elevations.

■ **LOCAL SITES** Common at Whistler, Harmony Lake, Piccolo Summit, Musical Bumps Trail, Blackcomb, Seventh Heaven, Rainbow Lake and Wedgemount Lake. Flowers mid-July through August and trickles into September.

41

SMALL-FLOWERED ALUMROOT

Heuchera micrantha • Saxifrage family: *Saxifragaceae*

■ DESCRIPTION Small-flowered alumroot is a perennial to
60 cm in height. Its small white flowers are abundant and
held on scapes (stems) up to 60 cm tall. The heart-shaped
leaves have long hairy stems and are basal. The leaves are
slightly longer than they are broad and distinguish this
plant from smooth alumroot (*H. glabra*), which has leaves
that are broader than they are long. The name "alumroot"
is given because the roots are very astringent.

■ HABITAT Wet cliff faces and stream banks at low
to high elevations.

■ LOCAL SITES Huge numbers in the Nairn
Falls area, trailside to Cheakamus and
Garibaldi lakes, and on damp rock faces
around Nita Lake. Flowers June to July.

FRINGECUP
Tellima grandiflora • Saxifrage family: *Saxifragaceae*

■ **DESCRIPTION** Fringecup is a perennial to 80 cm in height. Its fringed flowers are greenish, fragrant, 1 cm long and produced on 60- to 80-cm scapes (stems). The basal leaves are round to heart-shaped, deeply notched, 5–8 cm across; the scape leaves are smaller. When out of flower, fringecup can be confused with the piggy-back plant.

■ **HABITAT** Moist cool forests along the coast.

■ **NATIVE USE** The plants were crushed and boiled and the resultant infusion was used to treat sickness.

■ **LOCAL SITES** Large concentrations at Brandywine Falls, trailside to Cheakamus Lake and in lower Singing Pass. Flowers mid-May through June.

43

ONE-LEAFED FOAM FLOWER
Tiarella trifoliata var. unifoliata •

Saxifrage family: *Saxifragaceae*

■ **DESCRIPTION** One-leafed foam flower is a herbaceous perennial to 40 cm in height. Each wiry stem supports several tiny white flowers; the clusters of flowers are thought to resemble foam. The basal leaves are maple-shaped with 3–5 lobes.
■ **HABITAT** Shaded moist forests at low to mid elevations.
■ **LOCAL SITES** Common in the lower areas, trailside to Chekamus Lake, on lower trails to Garibaldi Lake and in lower Singing Pass. Flowers June through July.

TOLMIE'S SAXIFRAGE

Saxifraga tolmiei • Saxifrage family: *saxifragaceae*

■ **DESCRIPTION** Tolmie's saxifrage is a mat-forming perennial to 40 cm across. Its white flowers, to 1 cm across, have 5 petals with stamens between each; the stems, 5–10 cm tall, produce 1–4 flowers each. The small shiny leaves are succulent and slightly curled to prevent desiccation from strong sunlight.

■ **HABITAT** By the snowmelt line on rocky slopes at subalpine to alpine elevations.

■ **LOCAL SITES** Harmony Bowl, Piccolo Summit on Whistler Mountain. Flowers late July to August.

ALASKA SAXIFRAGE

Saxifraga ferruginea • Saxifrage family: *Saxifragaceae*

■ **DESCRIPTION** Alaska saxifrage is a perennial 20–40 cm in height. Its small white flowers to 1 cm across have 5 petals: the upper 3 petals are slightly broader and have 2 yellow dots. The leaves are basal, wedge-shaped and toothed above the middle. The genus name *Saxifraga* means "rock-breaking." It was thought that these plants could break rocks.

■ **HABITAT** Rock crevices, stream banks and seepage areas at low to high elevations.

■ **LOCAL SITES** Trailside above Brandywine Falls, Cougar Mountain and around Nairn Falls. Flowers May to July, depending on elevations.

46

PAINTBRUSH

Castilleja sp. • Figwort Family: *Scrophulariaceae*

■ **DESCRIPTION** The many species of paintbrush are difficult to distinguish. They range in height from 20 to 80 cm and there is frequent hybridization within their diverse growing range, making identification even harder. Paintbrush is a perennial with small lance-shaped leaves. Its actual flowers are small and inconspicuous — it is the showy red bracts that attract all the attention.

■ **HABITAT** Low-elevation grassy meadows and rocky outcrops to moist subalpine and alpine meadows.

■ **LOCAL SITES** Very common; subalpine clearings and meadows at Musical Bumps, Singing Pass, Seventh Heaven, Rainbow Lake, lower slopes on Blackcomb and One Mile Lake.

Flowers June to September.

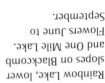

47

MOUNTAIN or ALPINE MONKEY-FLOWER

Mimulus tilingii • Figwort family: *Scrophulariaceae*

■ **DESCRIPTION** Alpine monkey-flower is a herbaceous perennial 5-20 cm in height. Its bright yellow flowers are almost duplicates of yellow monkey-flower (*M. guttatus*) except that they grow on very short leafy stems. It is most often seen as a beautiful low-forming mat in the Whistler alpine regions.

■ **HABITAT** Damp meadows and edges of cold streams at subalpine to alpine elevations.

■ **LOCAL SITES** Streamside between Symphony Bowl and Piccolo Summit. Full flower mid-August to mid-September.

PINK MONKEY-FLOWER or LEWIS' MONKEY-FLOWER

Mimulus lewisii • Figwort family: *Scrophulariaceae*

■ **DESCRIPTION** Lewis' monkey-flower is a clump-forming herbaceous perennial to 90 cm in height. Its flowers are snapdragon-shaped, 2–5 cm long and pinkish red with a yellow throat. The leaves are 7–10 cm long, opposite, and clasp the stems. When in flower, the clusters are a fabulous sight in the upper mountains. The species is named for Captain Meriwether Lewis of the Lewis and Clark Expedition.

■ **HABITAT** Cold stream edges and wet meadows from mid to subalpine elevations.

■ **LOCAL SITES** Large patches are found in moist meadows in upper Singing Pass. They supply a succession of pink flowers from July to August, a spectacular sight.

49

SLENDER BLUE PENSTEMON

Penstemon procerus • Figwort family: *Scrophulariaceae*

- **DESCRIPTION** Slender blue penstemon is an upright herbaceous perennial 10–60 cm in height. Its blue-purple flowers have the typical penstemon shape, funnel-like with a 5-lobed calyx; they are neatly arranged in whorls on the top portion of the stems. The basal leaves appear in tufts, while the stem leaves are opposite and lance-shaped.
- **HABITAT** Drier meadows and slopes at mid to alpine elevations.
- **LOCAL SITES** Common in the subalpine areas, Musical Bumps Trail, Seventh Heaven, Body Bag Bowl, Rainbow Lake. Full flower from the end of July through August.

50

DAVIDSON'S PENSTEMON
Penstemon davidsonii • Figwort family: *Scrophulariaceae*

■ **DESCRIPTION** Davidson's penstemon is an evergreen mat-forming shrub. Its showy flowers are unusually large, 2–3 cm long, blue-purple and produced in mass on stems 5–10 cm tall. The leaves are 1 cm long, opposite and evergreen. The species is named for Dr. George Davidson, an avid collector of western plants.

■ **HABITAT** Rocky slopes, cliffs at mid to high elevations.

■ **LOCAL SITES** Cliff faces across from Green Lake, draping over rocks at Nairn Falls, trailside to Wedgemount Lake, on low mounds along Musical Bumps Trail. Flowers June through August, depending on elevation.

51

WOOD BETONY or BRACTED LOUSEWORT

Pedicularis bracteosa • Figwort family: *Scrophulariaceae*

■ **DESCRIPTION** Bracted lousewort is a herbaceous perennial to 1 m in height. Its odd-looking flowers are pale yellow to pink-purple, shown off on 7–20-cm spikes. The green leaves are gracefully dissected to the point of resembling ferns, hence its other common name, fernleaf. This species is the tallest of the louseworts.

■ **HABITAT** Common in moist open forests at mid to high elevations.

■ **NATIVE USE** The leaf shapes were incorporated into the design of baskets.

■ **LOCAL SITES** Trailside in Singing Pass; thousands light up the sides of Seventh Heaven. Flowers end of July to August.

COW-PARSNIP or INDIAN CELERY

Heracleum lanatum • Carrot Family: *Apiaceae*

■ **DESCRIPTION** Cow-parsnip is a tall hollow-stemmed herbaceous perennial from 1 to 3 m in height. Its small white flowers are grouped in flat-topped umbrella-like terminal clusters to 25 cm across. It produces numerous small, egg-shaped seeds, 1 cm long, with a pleasant aroma. The large woolly compound leaves are divided into 3 leaflets, 1 terminal and 2 lateral (to 30 cm across). The genus name *Heracleum* is fitting for this plant of Herculean proportions. Giant cow-parsnip (*H. mantegazzianum*), an introduced species, grows to 4 m in height and can be seen in urban areas.

CAUTION: both species can cause severe blistering and rashes when handled.

■ **HABITAT** Moist forests, meadows, marshes and roadsides from low to high elevations.

■ **LOCAL SITES** Damp hills trailside to Cheakamus Lake, on slopes up Blackcomb Mountain and in seepage areas along the Sea to Sky Highway. Flowering starts end of June at lower areas, end of July at higher.

53

MOUNTAIN SWEET CICELY
Osmorhiza chilensis • Carrot family: *Apiaceae*

■ **DESCRIPTION** Mountain sweet cicely is a herbaceous perennial to 1 m in height. Its thin greenish flowers are small and hard to see in the forest; the thin seeds that develop are brown black and catch easily on socks and other clothing. The leaves are divided into threes, then threes again for a total of 9 leaflets. The licorice-scented root is reputed to have aphrodisiac powers.

■ **HABITAT** Cool moist forests at low to mid elevations.

■ **LOCAL SITES** Brandywine Falls, Nairn Falls, lower trails at Garibaldi Lake trailhead. Flowers end of May to June and sets seed very quickly.

HEAL-ALL or SELF-HEAL
Prunella vulgaris • Mint Family: *Lamiaceae*

■ **DESCRIPTION** Heal-all is an introduced herbaceous perennial to 40 cm in height. Its purple flowers are 2-lipped, 1–2 cm long, and borne in terminal spikes. The leaves are mostly lance-shaped, opposite, to 7 cm long. The stems are square. As its name suggests, heal-all has long been used medicinally. Seventeenth-century herbalist Nicholas Culpeper prescribed that it be "taken inwardly in syrups for inward wounds, outwardly in unguents and plasters for outward."

■ **HABITAT** Roadsides, forest edges, fields and parks at low to mid elevations.

■ **LOCAL SITES** Brandywine Falls, trailside to Rainbow Falls, lots around Nairn Falls to One Mile Lake. Flowers mid-June to August.

55

SUBALPINE BUTTERCUP or MOUNTAIN BUTTERCUP

Ranunculus eschscholtzii •

Buttercup Family: *Ranunculaceae*

■ **DESCRIPTION** Subalpine buttercup is a perennial 10-25 cm in height. Its shiny yellow flowers, to 3 cm across, have 5 petals and look as if they have been varnished: this sheen helps distinguish it from the fan-leafed cinquefoil (*Potentilla flabellifolia*). The buttercup's basal leaves are twice divided by 3 and forms a stalkless collar under the flowers. The species is named for Russian doctor and plant collector Johann F. Eschscholtz.

■ **HABITAT** Meadows, seepage areas, damp slopes at subalpine and alpine elevations.

■ **LOCAL SITES** Seventh Heaven, Blackcomb Lake, Decker Creek, Musical Bumps. Flowers July to August.

FLOWERS

RED COLUMBINE
Aquilegia formosa • Buttercup family: *Ranunculaceae*

■ **DESCRIPTION** Red columbine is a herbaceous perennial to 1 m in height. The drooping red-and-yellow flowers are up to 5 cm across and have 5 scarlet spurs arching backwards; they are almost translucent when the sun shines on them. The leaves are sea green above, paler below, to 8 cm across and twice divided by threes. In the head of the flower is a honey gland that can only be reached by hummingbirds and long-tongued butterflies. The hole that can sometimes be seen above this gland is caused by frustrated bumblebees chewing their way to the nectar. The name "columbine" means "dove," for the five arching spurs said to resemble five doves sitting around a dish.

■ **HABITAT** Moist open forests, meadows and creeksides at low to high elevations.

■ **LOCAL SITES** Damp meadows along Singing Pass, trailside to Cheakamus Lake, lots at mid to high elevations on Blackcomb. Flowers mid-June at lower elevations, mid-August at higher.

57

WESTERN ANEMONE or WESTERN PASQUE FLOWER
Anemone occidentalis • Buttercup Family: *Ranunculaceae*

■ **DESCRIPTION** Western anemone is a herbaceous perennial 15-30 cm tall when in flower and 30-60 cm when in seed. Its flowers are 5 cm across, creamy-white and often seen with blue-tinged sepals. The leaves are highly dissected and mainly basal, with a cluster of small leaves just under the flower. The seed heads are uniquely shaped like mop-tops and can be seen in the hundreds by mid- to late summer.

■ **HABITAT** Alpine and subalpine meadows.

■ **NATIVE USE** Some natives used an infusion of anemones to treat tuberculosis, but it is now considered poisonous, as is much of the buttercup family.

■ **LOCAL SITES** Common in the upper Whistler and Blackcomb areas. Starts flowering as soon as the snow disappears, from the end of June to July. Seed heads are dominant by the end of July to August.

BANEBERRY

Actaea rubra • Buttercup family: *Ranunculaceae*

■ **DESCRIPTION** Baneberry is a herbaceous perennial to 1 m in height. Its tiny white flowers are formed in rounded clusters, with protruding stamens that give them a fuzzy look. The ripened fruit are formed in elongated clusters of red or sometimes white berries. The crinkly leaves are coarsely toothed and divided 2 to 3 times into threes. CAUTION: the entire plant is considered poisonous.

■ **HABITAT** Moist cool forests at low to mid elevations.

■ **LOCAL SITES** Lower Singing Pass trailside to Cheakamus and Garibaldi lakes. Flowers at the beginning of June, with attractive berries by mid-August.

WESTERN MEADOWRUE

Thalictrum occidentale • Buttercup Family: *Ranunculaceae*

■ **DESCRIPTION** Western meadowrue is a graceful herbaceous perennial to 1 m in height. It is dioecious: the male flowers, to 2.5 cm long, have dangling purple green stamens and anthers; the females are greenish white, with 5-15 tiny pistils that mature into slender, pointed dry seeds. The blue-green leaves are divided 2 to 3 times into fan-shaped stalked leaflets.

■ **HABITAT** Most abundant at mid elevations in damp meadows and at forest edges.

■ **NATIVE USE** The seeds were crushed and rubbed into the hair and body as a perfume.

■ **LOCAL SITES** Spotty on Cougar Mountain, lots in damp meadows trailside to Cheakamus Lake. Flowers end of May to June.

WHITE MARSH-MARIGOLD

Caltha leptosepala • Buttercup family: *Ranunculaceae*

■ **DESCRIPTION** White marsh-marigold is a fleshy herbaceous perennial to 25 cm in height. Its attractive white flowers are borne 1 or 2 per stem, 2–4 cm across, with a greenish yellow centre. The rounded to heart-shaped leaves are mostly basal, and are longer than they are wide.
■ **HABITAT** Very happy to have its roots in frigid water at subalpine to alpine elevations.
■ **NATIVE USE** The leaves and buds were eaten raw or cooked (this is not recommended).
■ **LOCAL SITES** Flooded meadows at upper elevations to Rainbow and Blackcomb lakes. Full flower by mid-July.

PACIFIC BLEEDING HEART
Dicentra formosa • Bleeding heart family: *Fumariaceae*

■ **DESCRIPTION** B.C.'s native bleeding heart is very familiar, thanks to its resemblance to the many cultivated varieties. It is a herbaceous perennial to 40 cm in height, with pinkish heart-shaped flowers that hang in clusters of 5–15. The delicate fern-like leaves are basal, but sweep upwards so much that they almost hide the flowers. Under good growing conditions, bleeding hearts can cover hundreds of square metres. The species name *formosa* means "beautiful."

■ **HABITAT** Open broad-leafed forests with nutrient-rich topsoil.

■ **LOCAL SITES** Large quantities on damp hillsides around Cheakamus Lake, lower trails to Garibaldi and Rainbow lakes. Flowers mid-May to end of June, depending on elevation.

PINK CORYDALIS

Corydalis sempervirens •
Bleeding heart family: *Fumariaceae*

■ DESCRIPTION Pink corydalis is an annual/biennial to 60 cm in height. Its showy pink flowers grow to 3 cm long and have a splash of yellow on the tip. The bluish green leaves are alternate and multidivided; they warrant the plant in a woodland setting even without the flowers.

■ HABITAT Both dry and moist forests at low to mid elevations.

■ LOCAL SITES Nairn Falls, with heavy concentrations en route to One Mile Lake. Flowers mid-May to June.

64

FIREWEED
● *Epilobium angustifolium*
Evening primrose family: Onagraceae

■ **DESCRIPTION** Fireweed is a tall herbaceous perennial that reaches heights of 3 m in in good soil. Its purply red flowers grow on long showy terminal clusters. The leaves are alternate, lance-shaped like a willow's, 10–20 cm long and darker green above than below. The minute seeds are produced in pods 5–10 cm long and have silky hairs for easy wind dispersal. Fireweed flowers have long been a beekeeper's favourite. The name "fireweed" comes from the fact that it is one of the first plants to grow on burned sites; typically follows wildfires.

■ **HABITAT** Common throughout B.C. in open areas and burned sites.

■ **NATIVE USE** The stem fibres were twisted into twine and made into fishing nets, and the fluffy seeds were used in padding and weaving.

■ **LOCAL SITES** Common throughout the Whistler region. The lower to mid slopes of Blackcomb Mountain are a blaze of purple red flowers in July and August; flowers mid-June at lower elevations.

65

BROAD-LEAFED FIREWEED or ALPINE FIREWEED
Epilobium latifolium
• Evening primrose family: *Onagraceae*

■ **DESCRIPTION** Alpine fireweed is a showy herbaceous perennial to 40 cm in height. The large flowers are rose to purple and contrast well with the lanceolate bluish green leaves. The new shoots and young leaves can be used as a pot herb.

■ **HABITAT** Gravel bars along streams and creeks or on wet slopes at mid to high elevations.

■ **LOCAL SITES** Gravel stream bars along Piccolo Summit and moist slopes on Blackcomb below Seventh Heaven. Flowers July at mid levels, August at higher elevations.

BUNCHBERRY or DWARF DOGWOOD

Cornus canadensis • Dogwood family: *Cornaceae*

■ **DESCRIPTION** Bunchberry, a perennial no higher than 20 cm, is a reduced version of the Pacific dogwood tree (*C. nuttallii*). The tiny greenish flowers are surrounded by 4 showy white bracts, just like the flowers of the larger dogwood. The evergreen leaves, 4–7 cm long, grow in whorls of 5–7 and have parallel veins like the larger tree's. The beautiful red berries form in bunches (hence the name) just above the leaves in August. Bunchberry and Pacific dogwood have a habit of flowering twice, once in spring and again in late summer.

■ **HABITAT** From low to high elevations in cool moist coniferous forests and bogs.

■ **LOCAL SITES** Common at low to mid levels in the Whistler region. Can be seen carpeting the forest floor at Brandywine Falls; trailside to Wedgemount, Garibaldi, Cheakamus and Rainbow lakes. Flowers mid-May through July. Berries ripen mid-August.

SKUNK CABBAGE

Lysichiton americanum • Arum family: *Araceae*

■ **DESCRIPTION** Skunk cabbage is a herbaceous perennial to 1.5 m in height and as much as 2 m across. The small greenish flowers are densely packed on a fleshy spike and surrounded by a showy yellow spathe, the emergence of which is a sure sign that spring is near. The tropical-looking leaves can be over 1 m long and 50 cm wide.

■ **HABITAT** Common at low elevations in wet areas such as springs, swamps, seepage areas and floodplains.

■ **NATIVE USE** Skunk cabbage roots were cooked and eaten in spring in times of famine. It is said this poorly named plant has saved the lives of thousands.

■ **LOCAL SITES** Common around most small lakes and ponds. Extensive groves can be seen trailside to Rainbow Lake, often growing with lady fern (*Athyrium filix-femina*). Flowers May to June, depending on elevation.

STREAM VIOLET
Viola glabella • Violet family: *Violaceae*

■ DESCRIPTION Stream violet is a herbaceous woodland perennial to 25 cm in height. Its showy yellow flowers, to 2 cm across, each have 5 petals; the top 2 petals are pure yellow, while the bottom 3 have purple lines. The heart-shaped leaves are toothed and grow to 5 cm across. The flowers and leaves can be used in salads or steeped for tea.

■ HABITAT Needs a moist environment, open forests and meadows, at low to high elevations.

■ LOCAL SITES Like a yellow carpet trailside to Cheakamus Lake and Brandywine Falls. Lines the sides of Jordan Creek between Nita and Alpha lakes. Common at all elevations. Flowers mid-May at lower elevations, mid-July at higher.

WHITE-VEINED WINTERGREEN
or PAINTED WINTERGREEN
Pyrola picta • Heath family: *Ericaceae*

■ DESCRIPTION White-veined wintergreen is an attractive evergreen perennial to 30 cm in height. Its drooping flowers are yellowish green, with a waxy finish; they grow to 1 cm across and have a protruding, curved style. The basal leaves are 5-7 cm long and leathery, dark green with white venation. The species name *picta* means "painted," in reference to the beautiful leaves.

71

■ HABITAT Cool coniferous forests at low to mid elevations.

■ LOCAL SITES Lower-level trails to Garibaldi Lake, and from Nairn Falls to One Mile Lake. Flowers June and July.

PINK WINTERGREEN
Pyrola asarifolia • Heath family: *Ericaceae*

■ **DESCRIPTION** Pink wintergreen is an evergreen perennial to 40 cm in height. Its pinkish flowers, to 1 cm across, are carried on 30- to 40-cm stems. The leaves are roundish to elliptical, 5-8 cm long and formed in a basal rosette. *Pyrola* is from *pyrus* ("pear"), indicating that the leaves are sometimes pear-shaped.

■ **HABITAT** Moist forests with rich soil, where it can form extensive carpets 2 m by 8 m. Low to mid elevations.

■ **LOCAL SITES** Common; large concentrations trailside to Cheakamus and Rainbow lakes, Nairn Falls. Flowers June and July.

72

GREEN-FLOWERED WINTERGREEN
Pyrola chlorantha • Heath family: *Ericaceae*

■ DESCRIPTION Green-flowered wintergreen is an evergreen perennial to 25 cm in height. Its greenish yellow flowers are bowl-shaped, to 1 cm across, and have a protruding, curved style. The leaves are basal, roundish, and smaller than most wintergreens. This species is far more dainty than pink wintergreen (*P. asarifolia*; see page 73). *Chlorantha* means "green-flowered."

■ HABITAT Cool coniferous forests at low to mid elevations.

■ LOCAL SITES Common around the Garibaldi trail-head area and Nairn Falls. Flowers June and July.

ONE-SIDED WINTERGREEN
Orthilia secunda • Heath family: *Ericaceae*

■ DESCRIPTION One-sided wintergreen is an evergreen perennial to 20 cm in height. Its flowers, as the common name suggests, grow on one side of the stem; they are very small, white and bell-shaped, with a protruding, straight style. The leaves are mostly basal, toothed, oval and green. *Secunda* is from *secund*, meaning "one-sided."

■ HABITAT Cool coniferous forests at low to high elevations.

■ LOCAL SITES Common; lots trailside to Ancient Cedars on Cougar Mountain. Flowers June and July.

PRINCE'S PINE
Chimaphila umbellata • Heath family: *Ericaceae*

■ **DESCRIPTION** Prince's pine is a small evergreen shrub to 30 cm in height. The white to pink flowers are waxy, formed in loose nodding clusters held above the foliage; the resulting brownish seed capsules are erect and persist through the winter. The leathery leaves form in whorls, grow to 5 cm long and are sharply toothed.

■ **HABITAT** Cool coniferous forests at low to mid elevations.

■ **NATIVE USE** The leaves were steeped and used as a cold remedy.

■ **LOCAL SITES** Brandywine Falls, large patches on the Cheakamus, Rainbow and Garibaldi lakes trails. Flowers July to August.

MENZIES' PIPSISSEWA or LITTLE PRINCE'S PINE
Chimaphila menziesii • Heath family: *Ericaceae*

■ **DESCRIPTION** Little prince's pine is the daintier of the two *Chimaphila*, reaching a maximum height of only 15 cm. Its creamy white flowers are slightly fragrant, range from 1-3 per stem and nod above the foliage. The leaves are alternate, to 5 cm long, serrately edged and a darker green than the larger prince's pine (*C. umbellata*; see page 75). The species is named for Dr. Archibald Menzies, a surgeon and botanist who sailed with Captain George Vancouver.

■ **HABITAT** Cool coniferous forests at low to mid elevations. Both species are often found growing in the same area.

■ **LOCAL SITES** Trailside to Nairn Falls, One Mile Lake and Rainbow Falls. Flowers mid-June to July.

SITKA VALERIAN
Valeriana sitchensis • Valerian family: *Valerianaceae*

■ DESCRIPTION Sitka valerian is a herbaceous perennial to over 1 m in height. Its pinkish tubular flowers are slightly fragrant, grow in flat-topped clusters and are supported on long succulent stems. The leaves are thick, to 25 cm long and divided into 5-7 coarsely toothed segments. The roots are the source of the drug valerian, which is used as both a stimulant and antispasmodic.

■ HABITAT Moist mid-level stream banks, subalpine to alpine meadows and slopes.

■ LOCAL SITES Common at higher elevations; Harmony Bowl, Seventh Heaven and Rainbow Lake. Flowers June at mid elevations, July, August and September at higher.

BIRD'S-FOOT TREFOIL

Lotus corniculatus • Pea family: *Fabaceae*

■ **DESCRIPTION** Bird's-foot trefoil is an introduced perennial 10-40 cm in height. Its bright yellow flowers (3-15) are formed in compact rounded heads to 6 cm across. The leaves are divided into 3 top and 2 basal leaflets. It has been seeded on ski slopes to reduce soil erosion in summer months.

■ **HABITAT** Westward-facing slopes, fields and lawns at low to mid elevations.

■ **LOCAL SITES** Common in the Whistler-Blackcomb area. Flowers July and August.

ARCTIC LUPINE
Lupinus arcticus • Pea family: *Fabaceae*

■ DESCRIPTION Arctic lupine is a herbaceous perennial to 60 cm in height. Its pea-like flowers are light to dark blue and grow in a long terminal cluster; the resulting small seeds are contained in hairy pods to 4 cm long. The basal leaves are long-stemmed and have 6-8 leaflets with pointed tips. Arctic lupine is often seen in association with western anemone (*Anemone occidentalis*), partridgefoot (*Luetkea pectinata*), wood betony (*Pedicularis bracteosa*) and white-flowered rhododendron (*Rhododendron albiflorum*).

■ HABITAT Most abundant in open mid to subalpine elevations.

■ LOCAL SITES Common in the Whistler-Blackcomb region. Flowers July and August at high elevations.

SPREADING PHLOX
Phlox diffusa • Phlox family: *Polemoniaceae*

■ DESCRIPTION Spreading phlox is a mat-forming perennial that rarely exceeds 10 cm in height. Its star-shaped flowers, light pink to lavender and 1-2 cm across, bloom for weeks. The leaves are dark green, needle-like and fused in pairs. The genus name *Phlox* is Greek for "flame," an apt description of this plant in full flower.

■ HABITAT Common in the Whistler-Blackcomb region, from mid to high elevations on open talus slopes and exposed rock outcrops. Needs good drainage.

■ LOCAL SITES Musical Bumps Trail, Piccolo Summit and Flute Summit. Full flower by mid-August at high elevations.

SPREADING DOGBANE

Apocynum androsaemifolium •
Dogbane family: *Apocynaceae*

■ DESCRIPTION Spreading dogbane is a herbaceous perennial to 80 cm in height. Its pink flowers are bell-shaped, to 0.8 cm long, and hang in beautiful clusters at the branch ends. The egg-shaped leaves are in opposite pairs, to 8 cm long, and droop in hot temperatures. When broken, the red stems exude a milky juice.

■ HABITAT Exposed areas, dry forest edges and roadsides at low to high elevations.

■ NATIVE USE The stem fibres were used for cordage.

■ LOCAL SITES Nairn Falls to One Mile Lake along train tracks and beside the Sea to Sky Highway. Flowers June and July.

SIBERIAN MINER'S LETTUCE
Claytonia sibirica • Purslane family: *Portulacaceae*

■ **DESCRIPTION** Siberian miner's lettuce is a small annual to 30 cm in height. Its small white to pink flowers are 5-petalled and produced in abundance on long, thin, fleshy stems. The basal leaves are long-stemmed, opposite, ovate and, like the stems, succulent. Another species, *C. perfoliata*, differs in that its upper leaves are disc-shaped and fused to other flower stems. Siberian miner's lettuce was first discovered in Russia, where it was a staple food for miners. Early prospectors and settlers found both species made excellent early-season salad greens.

■ **HABITAT** Moist forest areas at low to mid elevations.

■ **LOCAL SITES** Common on the lower trails to Cheakamus and Garibaldi lakes. Flowers May through July.

RATTLESNAKE PLANTAIN
Goodyera oblongifolia • Orchid family: *Orchidaceae*

■ DESCRIPTION Rattlesnake plantain is an evergreen perennial to 40 cm in height. Its numerous small flowers are greenish white, orchid-shaped and produced on a spike 20-40 cm high; they have a tendency to grow on one side of the spike. The evergreen leaves are basal and rosette-like, from 5-10 cm long. They are criss-crossed by whitish veins, creating the rattlesnake pattern that gives the plant its common name.

83

■ HABITAT Usually found in dry to moist coniferous forests at low to mid elevations, with a moss-dominated understory.

■ LOCAL SITES Common at mid levels. Lots trailside to Garibaldi Lake, Nairn Falls, Brandywine Falls. Flowering starts mid-July, full flower displays by the beginning of August.

FAIRYSLIPPER

Calypso bulbosa • Orchid family: *Orchidaceae*

■ **DESCRIPTION** Fairyslipper is a delicate herbaceous perennial from corm to 20 cm in height. Its flower is mauve to light purple; the lower lip is lighter and decorated with spots, stripes and coloured hairs. The single leaf is broadly lanceolate and withers with the flower; a new leaf appears in late summer and remains through winter. This is the most beautiful of the native orchids.

■ **HABITAT** Mostly associated with Douglas fir and Western hemlock forests.

■ **NATIVE USE** The Haida boiled and ate the corms in small quantities; they have a rich, buttery flavour.

■ **LOCAL SITES** Hundreds decorate the lower trailsides to Garibaldi Lake and Nairn Falls. Flowers best seen mid-May to beginning of June.

84

HEART-LEAFED TWAYBLADE
Listera cordata • Orchid family: *Orchidaceae*

■ DESCRIPTION Heart-leafed twayblade is a single-stemmed herbaceous perennial to 20 cm in height. Its flowers have a forked bottom lip and range from pale green to purplish. Twayblades have 2 heart-shaped opposite leaves that grow midway up the stem; *cordata* refers to these heart-shaped leaves.

■ HABITAT Most often associated with moist coniferous forests at low to mid elevations.

■ LOCAL SITES
Lots on lower to mid trailsides to Garibaldi and Rainbow lakes. Flowers beginning of June.

ROUND-LEAFED REIN-ORCHID
Platanthera orbiculata • Orchid family: *Orchidaceae*

■ **DESCRIPTION** Round-leafed rein-orchid is a fleshy herbaceous perennial to 60 cm in height. Its showy flowers are greenish white, to 3 cm across. The stem has only 2 plump opposite-facing leaves, sitting on the ground. Round-leafed rein-orchid is an unmistakable treasure to find.

■ **HABITAT** Moist coniferous forests at low to mid elevations.

■ **LOCAL SITES** Patches can be seen in the Brandywine Falls area. Flowers mid-July to mid-August.

Platanthera stricta • Orchid family: *Orchidaceae*

■ **DESCRIPTION** Slender rein-orchid is one of the taller *Platanthera* species, growing to 75 cm in height. Its flowers are light green, without fragrance and openly spaced on the stem. The bottom leaves are broadly lanceolate and get progressively narrower up the stem. Slender rein-orchid is usually seen in large patches.

■ **HABITAT** Wet forests and meadows, seepage areas, and at mid elevations.

87

■ **LOCAL SITES** Extensive patches trailside to Cheakamus and Rainbow lakes, logging road to Singing Pass, mixed with white rein-orchid (*P. orbiculata*) on lower to mid-Blackcomb. Flowers June to July.

GOAT'S BEARD

Aruncus dioicus • Rose family: *Rosaceae*

■DESCRIPTION Goat's beard is a herbaceous perennial to 3 m in height. The plants are dioecious — male and female flowers appear on separate plants. The tiny white flowers are compacted into hanging panicles up to 60 cm long. The leaves are compound 3 times (thrice pinnate); leaflets are bright green with a toothed edge, tapering to a point. With a little imagination, the hanging flowers can look like a goat's beard.

90

■HABITAT Moist open woodlands, creeksides and wet rocky slopes at lower elevations.

■LOCAL SITES Large patches around Nita Lake and Nairn Falls. Common along the Sea to Sky Highway. Full bloom in mid-June.

KINNIKINNICK or BEARBERRY

Arctostaphylos uva-ursi • Heather family: *Ericaceae*

■**DESCRIPTION** Kinnikinnick is a trailing, mat-forming evergreen that rarely grows above 25 cm in height. Its fragrant, pinkish flowers bloom in spring and are replaced by bright red berries 1 cm across by late summer. The small, oval leaves grow to 3 cm long, are leathery and alternate. Grouse and bears feed on the berries.

■**HABITAT** Dry rocky outcrops and well-drained forest areas throughout B.C., from sea level to high elevations.

■**NATIVE USE**
Kinnikinnick is an eastern native word used to describe a tobacco mix. The leaves were dried and smoked, sometimes mixed with other plants.

■**LOCAL SITES** Upper dry sites around Brandywine Falls; rock outcrops and cliff faces around Alta Lake; exposed sites from Lost Lake to the alpine elevations. Can also be seen on rock outcrops along the BC Rail line between Brandywine Falls and Whistler Village.

SALAL

Gaultheria shallon • Heather family: *Ericaceae*

■DESCRIPTION Salal is a prostrate to mid-size bush that grows from 0.5 to 4 m in height. In spring the small pinkish flowers (1 cm long) hang like strings of tiny Chinese lanterns. The edible dark purple berries grow to 1 cm across and ripen by mid-August to September. Both the flowers and berries display themselves for several weeks. The dark green leaves are 7-10 cm long, tough and oval-shaped. Salal is often overlooked by berry pickers; the ripe berries taste excellent fresh and make fine preserves and wine.

■HABITAT Dry to moist forested areas along the entire coast.

■NATIVE USE Salal was an important food source for most native peoples. The berries were eaten fresh, mixed with other berries, or crushed and placed on skunk cabbage leaves to dry. The dried berry cakes were then rolled up and preserved for winter use.

■LOCAL SITES Common understory bush at Brandywine Falls; lower elevations at Whistler. Western tea-berry (*Gaultheria ovatifoloia*) is more common in Whistler. Flowering starts beginning of May. Fruit starts to ripen in August.

OREGON GRAPE

Mahonia nervosa • Barberry family: *Berberidaceae*

■DESCRIPTION Oregon grape is a smaller spreading understory shrub that is very noticeable when the bright yellow upright flowers are out. By midsummer the clusters of small green fruit (1 cm across) turn an attractive grape blue. The leaves are evergreen, holly-like, waxy and compound, with usually 9-17 leaflets. The bark is rough, light grey outside and brilliant yellow inside. Another species, tall Oregon grape (*M. aquifolium*), grows in a more open and dry location, is taller (2 m), and has fewer leaflets (5-9). The species name *aquifolium* means "holly-like."

■HABITAT Dry coniferous forests in southern coastal B.C. and Washington.

■NATIVE USE When steeped, the shredded stems of both species yield a yellow dye that was used in basket-making. The tart berries were usually mixed with sweeter berries for eating.

■LOCAL SITES Trailside to Cheakamus Lake, Garibaldi Lake and Nairn Falls. Flowers end of May. The berries begin to turn blue by August and persist through autumn.

OVAL-LEAFED BLUEBERRY
Vaccinium ovalifolium • Heather family: *Ericaceae*

■DESCRIPTION The oval-leafed blueberry is one of B.C.'s most recognized and harvested blueberries. It is a mid-size bush to 2 m in height. The pinkish bell-shaped flowers appear before the leaves and are followed by the classic blueberries. Rubbing the berries reveals a covering of dull bloom and a darker berry. The soft green leaves are smooth-edged, alternate, egg-shaped (no point) and grow to 4 cm in length. Fruit ripens as early as July at lower elevations and into September at higher levels.

■HABITAT Moist coniferous forests from sea level to high elevations.

■NATIVE USE Blueberries were a valuable and delicious food source. They were eaten fresh, mixed with other berries and dried for future use. As with all blueberries, they were also mashed to create a purple dye used to colour basket materials.

■LOCAL SITES Throughout the Whistler region. Alaskan blueberry (*V. alaskaense*) and black huckleberry (*V. membranaceum*) also grow here.

< *top, V. ovalifolium*
< *bottom, V. membranaceum*

SASKATOON BERRY or SERVICEBERRY
Amelanchier alnifolia • Rose family: *Rosaceae*

■**DESCRIPTION** Depending on growing conditions, the saskatoon berry can vary from a 1-m shrub to a small tree 7 m in height. The white showy flowers range from 1 to 3 cm across and often hang in pendulous clusters. The young reddish berries form early and by midsummer darken to a purple black up to 1 cm across. The light bluish green leaves are deciduous, oval-shaped and toothed above the middle.

■**HABITAT** Shorelines, rocky outcrops and open forests at low to mid elevations.

■**NATIVE USE** The berries were eaten fresh, mixed with other berries or dried for future use. On the great plains the berries were mashed with buffalo meat to make pemmican. The hard straight wood was a favourite for making arrows.

■**LOCAL SITES** Exposed sites around Brandywine Falls; large marble-size berries found on bushes dotting the pathway between Alpha, Nita and Alta lakes. Flowering starts May and the berries are fully ripe by mid-August.

THIMBLEBERRY
Rubus parviflorus • Rose family: *Rosaceae*

■ **DESCRIPTION** Thimbleberry is an unarmed shrub to 3 m in height. Its large white flowers open up to 5 cm across and are replaced by juicy bright red berries. The dome-shaped berries are 2 cm across and bear little resemblance to a thimble. The maple-shaped leaves grow up to 25 cm across and, when needed, make a good tissue substitute.

■ **HABITAT** Common in coastal B.C. and Washington in open forests at low to mid elevations.

■ **NATIVE USE** The large leaves were used to line cooking pits and cover baskets. The berries were eaten fresh, dried or mixed with other berries.

■ **LOCAL SITES** Common in moist forested sites from Brandywine Falls to the lower to mid elevations in Singing Pass and at all lower lakes. Flowering starts mid-May; the fruit matures at the end of July and into August.

Rubus sps. • *Rose family* • Rosaceae

■DESCRIPTION Of B.C.'s three blackberry species, only one is native to the region. The two introduced species require more sunshine to thrive. The three are easy to identify:

Trailing blackberry (*Rubus ursinus*) — the first to bloom (end of April) and set fruit (mid-July), it is often seen rambling over plants in and out of forested areas. The berries are delicious and the leaves can be steeped as a tea.

Himalayan blackberry (*R. discolor*) — this blackberry was introduced from India and has now taken over much of the Pacific Northwest. It is heavily armed, grows rampant to 10 m and is a prolific producer of berries. Blooming starts mid-June and the fruit sets by mid-August.

Cutleaf blackberry (*R. laciniatus*) — introduced from Europe, this berry is very similar to the Himalayan blackberry but less common.

■HABITAT Common on open wasteland, forest edges, roadsides and in ditches.

■LOCAL SITES Trailing blackberry is common at Brandywine Falls and in lower Singing Pass and the lower lakes area.

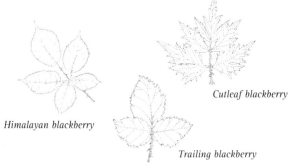

Cutleaf blackberry

Himalayan blackberry

Trailing blackberry

< *Ripe trailing blackberries*

SALMONBERRY

Rubus spectabilis • Rose family: *Rosaceae*

■DESCRIPTION Salmonberry is one of B.C.'s tallest native berry bushes. Though it averages 2-3 m, the bush can grow up to 4 m high. The pink bell-shaped flowers, 3-4 cm across, bloom at the end of February and are a welcome sight. Flowering continues until June, when both the flowers and ripe fruit can be seen on the same bush. The soft logan-shaped berries range in colour from yellow to orange to red, with the occasional dark purple. The leaves are compound, with 3 leaflets, much like the leaves of a raspberry. Weak prickles may be seen on the lower portion of the branches; the tops are unarmed. The berry's common name comes from its resemblance to the shape and colour of salmon eggs.

■HABITAT Common on the coast of B.C. in shaded, damp forests.

■NATIVE USE The high water content in the berries prevented them from being stored for any length of time. They were generally eaten shortly after harvesting.

■LOCAL SITES Good picking sites at Brandywine Falls and all lower lakes. Flowering starts in March, with the fruit ripening at the end of May to July.

BLACK RASPBERRY or BLACKCAP
Rubus leucodermis • Rose family: *Rosaceae*

■DESCRIPTION Black raspberry is a deciduous armed shrub to 2 m in height. Its white flowers are small, to 3 cm across and borne in terminal clusters of 3-7. The resulting fruit (1 cm across) starts off red but turns dark purple to black by July and August. The leaves have 3-5 leaflets supported on long, arching, well-armed stems. Black raspberries can be distinguished from other raspberries by the bloom, a whitish waxy coating, on the stems.

101

■HABITAT Open forests and edges at low to mid elevations.

■NATIVE USE The berries were eaten raw or dried into cakes for winter consumption.

■LOCAL SITES Trailside along Alta, Nita and Alpha lakes, Nairn Falls; patchy along the Cheakamus Lake Trail.

WOODLAND STRAWBERRY
Fragaria vesca • Rose family: *Rosaceae*

■DESCRIPTION Woodland strawberry is an unarmed herbaceous perennial to 20 cm in height. Its white flowers, 1-3 cm across, have 5 petals and a yellow centre. The delicious fruit, 1-3 cm across, is a small version of the cultivated strawberry. The leaves are compound, 3-5 cm across, and have 3 coarsely toothed leaflets. The blue-leaf or wild strawberry (*F. virginiana*) is similar to the woodland strawberry with 2 exceptions: the blue-leaf has bluish green leaves and the terminal teeth on its leaflets are shorter than the teeth on either side; the terminal teeth of the woodland strawberry are longer than the others.

■HABITAT Mainly in lower open forests, but can be found at mid to high elevations.

■NATIVE USE The juicy fruit was eaten fresh, the leaves were steeped for tea.

■LOCAL SITES
Common around Brandywine Falls. Flowers May to June. Blue-leaf strawberry can also be seen, but is less common.

BLACK GOOSEBERRY or BLACK SWAMP GOOSEBERRY

Ribes lacustre •
Currant and Gooseberry family: *Grossulariaceae*

■DESCRIPTION Black gooseberry is an armed shrub to 2 m in height. Its delicate reddish flowers are disc-shaped, to 0.5 cm across, and hang in drooping clusters of 7-15. The small, dark purple berries are bristly and hang in clusters of 3-4. The leaves are maple-shaped, with 5 lobes, from 2-5 cm across. The branches are covered with small golden spines, with larger spines at the nodes. Use caution when picking the berries: the spines can cause an allergic reaction in some people.

■HABITAT Moist open forests, lake edges at low to high elevations.

■NATIVE USE The berries were eaten by most groups.

■LOCAL SITES Common around the lower lakes (Nita, Alta, Alpha). Berries ripen July to August.

CREEPING RASPBERRY or FIVE-LEAFED BRAMBLE
Rubus pedatus • Rose family: *Rosaceae*

■DESCRIPTION Creeping raspberry is an unarmed trailing perennial from 20-100 cm long. Its white flowers to 2 cm across are produced singly. The small edible berries are glossy red and form in clusters of 1-5. The compound leaves have 5 coarsely toothed leaflets to 2.5 cm across.

■HABITAT Mostly found at mid levels in moist coniferous forests, but can be seen at low to subalpine elevations.

■NATIVE USE The berries were not widely used because they are small and difficult to gather.

■LOCAL SITES Common along Singing Pass, trailside to Rainbow, Cheakamus and Garibaldi lakes. Blankets the sides of trail to Wedgemount Lake. Flowers in June. Berries ripen August to September.

HIGHBUSH CRANBERRY or MOOSEBERRY

Viburnum edule • Honeysuckle family: *Caprifoliaceae*

■**DESCRIPTION** Highbush cranberry is usually seen as a straggling bush 0.5-3 m in height. Its small white flowers to 1 cm across are borne in rounded clusters nestled between the paired leaves. The resulting red berries, to 1.5 cm across, grow in clusters of 2-5. The leaves are mostly 3-lobed and opposite.

■**HABITAT** Open forests and edges at low to mid elevations.

■**NATIVE USE** The tart berries were generally preserved for several months before eating.

■**LOCAL SITES** Trailside to Brandywine Falls, Rainbow Lake, Singing Pass, lower slopes of Blackcomb Mountain; lower lakes around Whistler Village. Flowers end of May to June. Berries ripen in September and last through October.

MAIDENHAIR FERN

Adiantum pedatum • Polypody family: *Polypodiaceae*

■**DESCRIPTION** Maidenhair fern is a delicate-looking fern with an almost tropical appearance. The fan-shaped fronds carry the dainty green leaflets (pinnules), which contrast well with the dark stems (stipes) that grow up to 60 cm in length. The reproducing sori under the pinnules are visible in late summer and fall. The genus name *Adiantum*, meaning "unwetted," refers to the way the fronds repel water.

■**HABITAT** Moist cliff faces at low to mid elevations.

107

■**NATIVE USE** The shiny black stipes were used in basket-making.

■**LOCAL SITES** Patches along trail to Garibaldi Lake.

LADY FERN
Athyrium filix-femina • Polypody family: *Polypodiaceae*

■DESCRIPTION Lady fern is a tall fragile fern to 2 m in height. The apple green fronds average up to 30 cm across and are widest below the centre, tapering at top and bottom. This diamond shape distinguishes the lady fern from the similar-looking spiny wood fern (*Dryopteris expansa*; see page 111) whose fronds have an abrupt triangular form. The fronds die off in winter and emerge again in April. The horseshoe-shaped sori appear on the back of the fronds in spring.

■HABITAT Moist forests with nutrient-rich soils.

■NATIVE USE The young fronds (fiddleheads) were sometimes eaten in April.

■LOCAL SITES Common; trailside to Rainbow Lake. Groves often mixed with skunk cabbage (*Lysichiton americanum*; see page 69).

108

DEER FERN

Blechnum spicant • Polypody family: *Polypodiaceae*

■DESCRIPTION Deer fern can be distinguished from licorice fern (*Polypodium glycyrrhiza*) and sword fern (*Polystichum munitum*; see page 112) by its two distinct types of frond, sterile and fertile. The sterile fronds grow up to 75 cm long, are tapered at both ends and usually lie flat. The fertile or spore-producing fronds are erect from the centre of the plant and can grow up to 75 cm in height. Deer ferns are good winter browse for deer.

■HABITAT Moist forested areas with plenty of rainfall.

■LOCAL SITES Along lower trails to Garibaldi Lake.

PARSLEY FERN or MOUNTAIN FERN
Cryptogramma crispa • Polypody family: *Polypodiaceae*

■DESCRIPTION Parsley fern is a small evergreen to 30 cm in height. It has two sets of fronds, sterile and fertile (spore-producing). The sterile fronds are evergreen, deeply dissected, parsley-like and grow to 20 cm in height. The fertile fronds have less-congested foliage, their leaf margins are rolled and cover the sori, and they grow to 30 cm in height. The species name *crispa* refers to the crisped look of the fronds.

110

■HABITAT Dry sites, typically rocky outcrops or slopes at low to high elevations.
■LOCAL SITES Upper trails to Cal-Cheak Suspension Bridge; mid elevations in Singing Pass; at the viewing platforms at Nairn Falls.

SPINY WOOD FERN or SHIELD FERN
Dryopteris expansa • Polypody family: *Polypodiaceae*

■**DESCRIPTION** Spiny wood fern is an elegant plant to
1.5 m tall. The pale green fronds are triangular in shape,
average up to 25 cm across and die off in winter. In spring
the rounded sori are produced on the underside of the
fronds. Spiny wood fern is similar in appearance and
requirements to lady fern (*Athryium filix-femina*).
■**HABITAT** Common in moist forests at low to mid
elevations.

111

■**LOCAL SITES** Lower trailsides to Garibaldi, Wedgemount
and Rainbow lakes.

WESTERN SWORD FERN
Polystichum munitum • Polypody family: *Polypodiaceae*

■**DESCRIPTION** Western sword fern is southern B.C.'s most common fern. It is evergreen and can grow to 1.5 m in height. The fronds are dark green with side leaves (pinnae) that are sharply pointed and toothed. On the underside of the fronds a double row of sori forms midsummer and turns orange by autumn. The fronds are in high demand in eastern Canada for floral decorations. The species name *munitum* means "armed," referring to the side leaves that resemble swords.

■**HABITAT** Dry to moist forest at lower elevations near the coast, where it can form pure groves.

■**NATIVE USE** The ferns were used to line steaming pits and baskets, and were placed on floors as sleeping mats.

■**LOCAL SITES** Brandywine Falls, lower lakes around Whistler Village and lower Garibaldi Lake Trail.

112

OAK FERN
Gymnocarpium dryopteris •
Polypody family: *Polypodiaceae*

■DESCRIPTION Oak fern is a deciduous plant to 40 cm in height. Usually seen in patches, the fragile-looking leaves are triangular and appear to be in 3 segments. The stipe or stem is golden brown.

■HABITAT Moist forests at low to high elevations.

■LOCAL SITES Common at mid elevations; like a soft green carpet covering hundreds of square metres on the slopes toward Wedgemount Lake. Large patches on Singing Pass Trail after the Russet Lake turnoff.

COMMON HORSETAIL

Equisetum arvense • Horsetail family: *Equisetaceae*

■DESCRIPTION Common horsetail is a herbaceous perennial to 75 cm in height. It has two types of stems, fertile and sterile, both hollow except at the nodes. The fertile stems are unbranched, to 30 cm in height, and lack chlorophyll; they bear spores in the terminal head. The green sterile stems grow to 75 cm in height and have leaves whorled at the joints. Horsetails are all that is left of a prehistoric family, some members of which grew to the size of trees.

■HABITAT Low wet seepage areas, meadows, damp sandy soils and gravel roads from low to high elevations.

■LOCAL SITES Extensive patches around Alpha Lake, on road to Singing Pass, lower trailside to Rainbow Lake, wet slopes on lower Blackcomb.

ROUND-LEAFED SUNDEW
Drosera rotundifolia • Sundew family: *Droseraceae*

■DESCRIPTION There are about 100 species of sundew around the world, and all of them eat insects. B.C.'s native round-leafed sundew (below, right) is a small perennial, 5-25 cm high, with inconspicuous white flowers. It is the leaves that make this plant a curiosity. They are equipped with fine red hairs, each tipped with a shiny globe of reddish secretion. Small insects are attracted to the secretion and get stuck in it; the leaf then slowly folds over and smothers the unsuspecting visitors. The plant's favourite foods are mosquitoes, gnats and midges.

■HABITAT Peat bogs throughout the west coast of B.C.

■NATIVE USE The whole plant is acrid, and the leaves were once used to remove corns, warts and bunions.

■LOCAL SITES Can be seen growing with the narrow-leafed great sundew (*D. anglica*; below, left) around the ponds above Brandywine Falls. Both species carpet the ground.

CAT-TAIL
Typha latifolia • Cat-tail family: *Typhaceae*

■DESCRIPTION Cat-tails are semi-aquatic perennials that can grow to 2.5 m in height. The distinctive "tail," a brown spike, is 15-20 cm long and 3 cm wide and made up of male and female flowers. The lighter-coloured male flowers grow at the top and usually fall off, leaving a bare spike above the familiar brown female flowers. The sword-shaped leaves are alternate and spongy at the base.

■HABITAT Common in B.C. at low to mid elevations, at lakesides and riversides and in ponds, marshes and ditches.

■NATIVE USE The long leaves were used to weave mats and the fluffy seeds to stuff pillows and mattresses.

■LOCAL SITES Lost Lake, Alpha Lake and small ponds between the two lakes.

RUNNING CLUBMOSS
Lycopodium clavatum • Clubmoss family: *Lycopodiaceae*

■**DESCRIPTION** Running clubmoss is a curious creeping evergreen that looks like it is made of bright green pipe cleaners. Like all clubmosses, it has no flowers and reproduces by spores. These are held in terminal cones on vertical stalks to 25 cm in height. The evergreen leaves are lance-shaped and arranged spirally around the stem. Running clubmoss grows horizontally across the ground, with irregular rooting. The spores are used medicinally and in industry.

■**HABITAT** Dry to moist coniferous forests at low to high elevations.
■**LOCAL SITES** Trailside from Brandywine Falls to Cal-Cheak Suspension Bridge; Garibaldi, Showh and Rainbow lakes. The largest patches are found trailside to Wedgemount Lake.

GROUND CEDAR

Lycopodium complanatum •
Clubmoss family: *Lycopodiaceae*

■DESCRIPTION Appropriately named, ground cedar looks as if someone has poked small western red cedar branches into the ground. This is one of 3 species of clubmoss found in the Whistler area. Its upright stems grow to 30 cm in height and are formed on horizontal stems that creep along the ground just below the soil surface.

■HABITAT Dry to moist coniferous forests at low to mid elevations.

■LOCAL SITES Elevated trails in the Brandywine Falls area.

BUCKBEAN or BOGBEAN
Menyanthes trifoliata •
Buckbean family: *Menyanthaceae*

■DESCRIPTION Buckbean is an aquatic perennial to 30 cm above the water. Its white flowers have 5 petals and straggly, protruding hairs; their foul smell attracts various insects for pollination. The succulent leaves are basally clustered and divided into 3 leaflets, hence the species name.

■HABITAT Ponds and lakes at low to mid elevations.

■LOCAL SITES Common in lower lakes and ponds; dominant in ponds between Brandywine Falls and Cal-Cheak Suspension Bridge. Flowers May to June.

NARROW-LEAFED COTTON GRASS
Eriophorum angustifolium • Sedge family: *Cyperaceae*

■DESCRIPTION Cotton grass is a rhizomatous herbaceous perennial to 70 cm in height. The inconspicuous flowers are held on triangular stems 30-70 cm long. When mature, the flowers are covered with silky white hairs (cotton) to 3 cm long. The flat leaves look like grass. The species name *angustifolium* means "narrow-leafed."
■HABITAT Peat bogs at low to high elevations; tolerates shallow water.
■LOCAL SITES Dense colonies in ponds above Brandywine Falls. Seeds June to July.

123

PLANTAIN
Plantago spp. • Plantain family: *Plantaginaceae*

■**DESCRIPTION** Ribwort plantain (*P. lanceolata*; below, left) and common plantain (*P. major*; below, right) are introduced perennials that have made weeds of themselves worldwide. Ribwort grows to 60 cm in height, has curious flowers and long, basal, lance-shaped leaves. Common plantain grows to 40 cm in height and has tight spikes of green flowers and large oval leaves.

■**HABITAT** Both species invade lawns and roadsides at low to mid elevations.

■**LOCAL SITES** Common plantain is abundant, often found on roadsides.

WILD SARSAPARILLA
Aralia nudicaulis • Ginseng family: *Araliaceae*

■DESCRIPTION Wild sarsaparilla is a herbaceous perennial to 40 cm in height. The easily overlooked flowers are greenish white, 5-petalled and held in small rounded clusters; they are replaced by small clusters of dark purple berries in August. Sarsaparilla produces from its rhizomes a central stem that has 3 compound leaves divided again into 3-5 leaflets.

■HABITAT Moist forests at low to mid elevations.

■NATIVE USE The rhizomes were often steeped to make tea.

■LOCAL SITES Common from Nairn Falls to One Mile Lake. Flowers in June; ripe seed by August.

125

YELLOW POND LILY

Nuphar polysepalum • Water lily family: *Nymphaeaceae*

■ **DESCRIPTION** Yellow pond lily is a long-stemmed aquatic perennial. Its striking yellow flowers to 10 cm across are a familiar summer sight in lakes and ponds. A large round stigma dominates the centre of these large, waxy flowers. The heart-shaped floating leaves, or pads, grow to 40 cm long. The huge rhizomes when exposed at low water levels are sought after by bears. The genus name *Nuphar* means water lily.

■ **HABITAT** Ponds, lakes, marshes at low to mid elevations.

■ **NATIVE USE** The seeds, called "wok as," were gathered and used as a food source.

■ **LOCAL SITES** Small ponds between Nita and Alpha lakes; between Brandywine Falls and Cal-Cheak Suspension Bridge. Flowering begins in May and continues through the summer.

PINESAP
Hypopitys monotropa •
Indian pipe family: *Monotropaceae*

■DESCRIPTION Pinesap is a curious-looking fleshy saphrophyte to 30 cm in height. Its flowers and stem are yellowish pink. Approximately 5-10 flowers grow on only one side of the curved stem; they start off pointing downward but, as they mature, the colour darkens and they turn to face upward. The species name *monotropa* means "one-sided."

■HABITAT Moist coniferous forests at low to mid elevations.

■LOCAL SITES A beautiful display can be seen under the dwarf pine forest between Brandywine Falls and Cal-Cheak Suspension Bridge.

128

LEATHERLEAF SAXIFRAGE

Leptarrhena pyrolifolia • Saxifrage family: *Saxifragaceae*

■DESCRIPTION Leatherleaf saxifrage is a moisture-loving perennial 15-30 cm in height. Its tiny white flowers are held in round terminal clusters. The leathery leaves, to 7 cm long, are dark green on top and a lighter green underneath. As cold weather comes to the meadows, the stems and seed heads turn brilliant red.

■HABITAT Wet meadows at subalpine elevations.

■LOCAL SITES Meadows around Rainbow Lake and Harmony Meadows. Flowers July to August. Red colouring mid-September to October.

WILD GINGER
Asarum caudatum • Birthwort family: *Aristolochiaceae*

■DESCRIPTION Wild ginger is a trailing evergreen perennial that forms patches several metres wide. Its bell-shaped solitary flower is purplish brown, to 5 cm across, with 3 pointed lobes. The heart-shaped leaves, to 10 cm across, are formed in opposite pairs in the nodes. The whole plant has a mild ginger fragrance when crushed.

■HABITAT Moist shaded forests with rich humus, at low to mid elevations.

■NATIVE USE The scented plants were put in bathwater and the roots were boiled and drunk as a tea to ease stomach problems.

■LOCAL SITES Like a ground cover trailside to Cheakamus Lake; lots at Brandywine Falls, Nairn Falls, and on nature trail along Blackcomb Creek to Lost Lake. Flowers mid-May to August.

NOOTKA ROSE

Rosa nutkana • Rose family: *Rosaceae*

■DESCRIPTION The largest of B.C.'s native roses, the Nootka rose grows to 3 m in height. The showy pink flowers are 5-petalled, fragrant, 5 cm across and usually solitary. The compound leaves have 5-7 toothed leaflets and are armed with a pair of prickles underneath. The reddish hips are round and plump, 1-2 cm across, and contrast well with the dark green foliage.

■HABITAT Open low-elevation forests throughout B.C.

■NATIVE USE Rosehips were strung together to make necklaces and the flowers were pressed to make perfume. Rosehips were only eaten in times of famine.

■LOCAL SITES Trailside from Cal-Cheak Suspension Bridge to Brandywine Falls. Flowers in June, with the hips developing colour in August.

BALDHIP or WOODLAND ROSE
Rosa gymnocarpa • Rose family: *Rosaceae*

■**DESCRIPTION** The baldhip rose is B.C.'s smallest native rose. It is often prostrate, to 1.4 m in height. The tiny pink flowers are 5-petalled, delicately fragrant, 1-2 cm across and usually solitary. The compound leaves are smaller than the Nootka rose and have 5-9 toothed leaflets. The spindly stems are mostly armed with weak prickles. A good identifier is this rose's unusual habit of losing its sepals, leaving the hip bald — hence the species name *gymnocarpa*, which means "naked fruit." Rosehips have a higher concentration of vitamin C than oranges and make an excellent jelly or marmalade.

■**HABITAT** Dry open forests at lower elevations, from southern B.C. to the redwood forests of California.

■**LOCAL SITES** Upper levels of Brandywine Falls and Nairn Falls. Flowers June.

HARDHACK or STEEPLEBUSH
Spiraea douglasii • Rose family: *Rosaceae*

■DESCRIPTION Hardhack is an upright deciduous bush to 2 m in height. Its tiny pink flowers group together to form fuzzy pyramidal clusters up to 15 cm tall. The resulting brown fruiting clusters persist on the bush through winter. The alternate leaves are elliptic to oval, toothed above the middle, 5-10 cm long, dark green above and a felty paler green below.

■HABITAT Prefers moist conditions, can be seen growing in ditches and bogs and at lakesides from low levels to subalpine meadows.

■NATIVE USE The tough wiry branches were used to make halibut hooks, scrapers and hooks for drying and smoking salmon.

■LOCAL SITES Thickets along boardwalk at One Mile Lake and Alpha Lake; common along the Sea to Sky Highway. In full flower mid-June at Nairn Falls, mid-July at higher elevations.

NINEBARK
Physocarpus capitatus • Rose family: *Rosaceae*

■DESCRIPTION Ninebark is an upright deciduous shrub to 4 m in height. Its tiny white flowers are grouped into rounded clusters to 7 cm across. The fruit are reddish brown inflated seed capsules. The maple-shaped leaves are 3-5 lobed, shiny green above, paler below, to 7 cm long. It is debatable whether the shaggy bark has nine layers. The species name *capitatus* refers to the rounded heads of the flowers.

134

■HABITAT Usually seen on moist sites, in open forests and at streams and lakesides, but also on dry rocky areas at lower elevations.

■NATIVE USE Ninebark was used medicinally.

■LOCAL SITES Between Nairn Falls and One Mile Lake and at Brandywine Falls. Flowers mid- to late July.

RED-BERRIED ELDER or RED ELDERBERRY
Sambucus racemosa • Honeysuckle family: *Caprifoliaceae*

■DESCRIPTION Red-berried elder is a bushy shrub to 6 m in height. Its small flowers are creamy-white and grow in pyramidal clusters 10-20 cm long. The berries that replace them take up to 3 months to turn bright red; they are considered poisonous to people when eaten raw but are a favourite food for birds. The leaves are compound, 5-15 cm long, with 5-9 opposite leaflets.

CAUTION: the berries are considered poisonous.

135

■HABITAT Moist coastal forest edges and roadsides. The blue-berried elder (*S. caerulea*) is found more in the Interior and the Gulf Islands.

■NATIVE USE The pithy branches were hollowed out and used as blowguns.

■LOCAL SITES Brandywine Falls to the lower levels of Whistler. In flower mid-July on Blackcomb Mountain, compared with mid-April at sea level.

FALSE AZALEA or FOOL'S HUCKLEBERRY
Menziesia ferruginea • Heather family: *Ericaceae*

■DESCRIPTION False azalea is an upright, deciduous shrub to 3 m in height. Its flowers, which resemble a huckleberry's, are a dull copper colour, bell-shaped, to 8 mm long, with long stems. The small fruit (5 mm long) is a dry, four-valved capsule that is not edible. The leaves are elliptic, bluish green on top, whitish green below, 3–6 cm long. They appear to grow in whorls. The genus name *Menziesia* is after Archibald Menzies, a naval surgeon and botanist who sailed with Captain George Vancouver and collected plants on the West Coast.

■HABITAT Moist forested sites, especially in wetter areas, at low to high elevations.

■LOCAL SITES Large specimens at Brandywine Falls, lower Singing Pass, Alta and Nita lakes, Cougar Mountain and trailside to Wedgemount Lake. Flowering begins mid-May to June.

WESTERN BOG LAUREL

Kalmia microphylla ssp. *occidentalis* • Heather family: *Ericaceae*

■ **DESCRIPTION** B.C.'s native laurel is a small lanky evergreen no more than 60 cm high. Its beautiful pink flowers (2.5 cm across) have a built-in pollen dispenser. A close look reveals that some of the stamens are bent over; these spring up when the flower is disturbed, dusting the intruder with pollen. The leaves are opposite, lance-shaped, 2-4 cm long, shiny dark green above, felty white below, with edges strongly rolled over. The plant is poisonous and should not be confused with Labrador tea (*Ledum groenlandicum*; see page 140), which it resembles from above.

CAUTION: the plant is poisonous.

■ **HABITAT** Peat bogs and lakeshores at low to high elevations throughout B.C.

■ **NATIVE USE** The leaves were boiled and used in small doses for medicinal purposes.

■ **LOCAL SITES** Pond edges above Brandywine Falls, mixed with Labrador tea (*Ledum groenlandicum*), wet meadows trailside to Rainbow and Blackcomb lakes. June and July are good times to find this beauty in flower at mid to high elevations.

LABRADOR TEA
Ledum groenlandicum • Heather family: *Ericaceae*

■**DESCRIPTION** Most of the year, Labrador tea is a gangly small shrub to 1.4 m in height. In spring the masses of small white flowers turn it into the Cinderella of the bog. The evergreen leaves are lance-shaped, alternate, 4-6 cm long, with the edges rolled over. The leaves can be distinguished from those of the poisonous bog laurel (*Kalmia microphylla*; see page 139) by their flat green colour on top and rusty-coloured hairs beneath. To be safe, pick the leaves only when the shrub is in flower.

■**HABITAT** Peat bogs, lakesides and permanent wet meadows, low to alpine elevations.

■**NATIVE USE** The leaves have long been used by native groups across Canada as infusions. Early explorers and settlers quickly picked up on this caffeine substitute. Caution must be taken – not all people can drink it.

■**LOCAL SITES** Good concentrations around ponds between Brandywine Falls and Cal-Cheak Suspension Bridge, Nita and Alta lakes. Flowers in June.

PINK MOUNTAIN HEATHER
Phyllodoce empetriformis • Heather family: *Ericaceae*

■**DESCRIPTION** Pink mountain heather is a mat-forming evergreen shrub 10-50 cm in height. Its rose-pink flowers are bell-shaped to 1 cm long and are held out on long, slender stalks. With their stiff needle-like leaves to 1 cm long, the branches resemble miniature conifers. A hike into alpine meadows is well worth the effort to see this beautiful fragrant plant.

■**HABITAT** Rocky sites at subalpine to alpine elevations.

■**LOCAL SITES** Common; when in flower, from July to August, its pink flowers blanket vast exposed areas on Whistler, Blackcomb and Garibaldi mountains. Yellow mountain heather, *P. granduliflora*, can be seen toward Blackcomb Lake.

WHITE MOUNTAIN HEATHER
Cassiope mertensiana • Heather family: *Ericaceae*

■**DESCRIPTION** White mountain heather is a low-growing shrub to 30 cm in height and up to several metres wide. Its nodding flowers are bell-shaped, pure white, and grow singularly on small stalks arising from the leaf axils. The tiny evergreen leaves are scale-like and arranged in 4 rows.
■**HABITAT** Common at subalpine and alpine elevations.
■**LOCAL SITES** Common; can be seen growing with pink mountain heather (*Phyllodoce empetriformis*; see page 141). Flowers July to August.

WHITE-FLOWERED RHODODENDRON
Rhododendron albiflorum • Heather family: *Ericaceae*

■**DESCRIPTION** White-flowered rhododendron is a deciduous bush to 2 m in height. Its showy flowers are white cream, to 2 cm long, and grow in clusters of 2 to 4 along the stems. The leaves are alternate, from 3-7 cm long, and the upper surface is shiny but also slightly hairy.

■**HABITAT** Moist slopes in the open or on the edges of coniferous forests at subalpine elevations.

■**NATIVE USE** The leaves were used for tea by some groups; others boiled the buds to use as a sore throat remedy or chewed them to treat stomach ulcers.

143

■**LOCAL SITES** Trailside to Rainbow Lake, dominant on the descent to Harmony Lake from the Roundhouse, Whistler, Seventh Heaven and Singing Pass. Can be seen in thickets under the chairlift to the Roundhouse. Flowers mid-July to beginning of August.

RED-FLOWERING CURRANT

Ribes sanguineum • Currant family: *Grossulariaceae*

■DESCRIPTION Red-flowering currant is a deciduous upright bush from 1.5-3 m in height. Its flowers range in colour from pale pink to bright crimson and hang in 8-12 cm panicles. The bluish black berries (1 cm across) are inviting to eat but are usually dry and insipid. The leaves are 5-10 cm across, maple-shaped, with 3-5 lobes. Currants and gooseberries are in the same genus (*Ribes*); a distinguishing feature is that currants have no prickles, while gooseberries do.

■HABITAT Dry open coastal forests at low to mid elevations.

■LOCAL SITES Can be seen around Whistler Village. Flowers mid-May to June.

Oplopanax horridus • Ginseng family: *Araliaceae*

■DESCRIPTION Devil's club is a deciduous shrub from
1-3 m in height. Its small white flowers are densely packed
into pyramidal clusters approximately 15 cm long. The
flowers bloom in May and are replaced by showy scarlet
berries in August; these are not considered edible. The large
leaves are maple-like, alternate, to 30 cm across, with
spines in the larger veins on both sides. The stems are
sprawling, awkward-looking and very well-armed with
spines to 1 cm long. The species name *horridus* comes to
mind when you accidentally encounter this shrub.

■HABITAT Moist forested areas with rich soil at low to mid
elevations.

■NATIVE USE Next to hellebore, devil's club was coastal
natives' most valued medicinal plant. Infusions and poultices
were used to relieve arthritis, fevers, colds and infections.

■LOCAL SITES Common; trailside to Ancient Cedars Grove
on Cougar Mountain, Brandywine Falls, Garibaldi Lake;
forms impenetrable thickets to Cheakamus Lake and
Rainbow Lake. Red berries by August to September.

RED-OSIER DOGWOOD
Cornus stolonifera • Dogwood family: *Cornaceae*

■**DESCRIPTION** Red-osier dogwood is a mid-size deciduous shrub to 5 m in height. Its small white flowers (4 mm across) are grouped together to form dense round clusters approximately 10 cm across. By August they have been replaced by bunches of dull white inedible berries to 8 mm across. The leaves are typically dogwood: opposite, to 10 cm long, with parallel veins. Younger branches are pliable and have an attractive red colour.

■**HABITAT** Moist to wet areas, usually forested, at low to mid elevations.

■**NATIVE USE** The small branches were used for weaving, barbecue racks, fuel for smoking salmon and latticework for fishing weirs.

■**LOCAL SITES** Common; very noticeable mid-June as its abundant flowers decorate the sides of the Sea to Sky Highway from Brandywine Falls to Whistler Village. Berries ripen late August.

FALSEBOX or MOUNTAIN LOVER
Pachistima myrsinites • Staff tree family: *Celastraceae*

■DESCRIPTION Falsebox is a small evergreen shrub to 75 cm in height. Its tiny maroon flowers go unnoticed by all but a curious few. The evergreen leaves are elliptic, leathery, tooth-edged, to 3 cm long. Falsebox is an attractive bush more noticed for its foliage than its flowers. The species name *myrsinites* is Greek for myrrh, in reference to the fragrant flowers.

■HABITAT Forested mountain slopes at low to mid elevations. Rare on the coast, but abundant in the B.C. interior.

■LOCAL SITES Common at mid levels around Whistler Mountain; good selection along logging road to Singing Pass; Cougar Mountain, Garibaldi Lake. Flowers May to June onward.

SWEET GALE

Myrica gale • Wax myrtle family: *Myricaceae*

■DESCRIPTION Sweet gale is an aromatic deciduous shrub to 1-3 m in height. Its flowers are displayed in yellowish male and female catkins, which appear in spring before the leaves. The fruit are tiny brown cone-like husks that persist through the winter. The thin leaves are flat green above and whitish below, coarsely toothed above the middle, to 5 cm long. The genus name *Myrica* means "perfume," a reference to the sweet-scented leaves.

■HABITAT Prefers to have its feet wet in shallow fresh water: ponds, lakes and swamps at low to mid elevations.

■LOCAL SITES Lakeside at Lost, Alta, Nita and One Mile lakes; lots in small ponds between Nita and Alta lakes.

SITKA MOUNTAIN ASH
Sorbus sitchensis • Rose family: *Rosaceae*

■DESCRIPTION Sitka mountain ash is a small, multi-stemmed bush or thicket from 1.5 to 4.5 m in height. Its compound bluish green leaves have 7-13 leaflets, 11 being the norm. The tiny white flowers are in terminal clusters, 5-10 cm across. In August and September the bushes and trees display a wonderful show of bright red-orange berries. The bark is thin and shiny grey. The native mountain ash should not be confused with the larger European mountain ash (*S. Aucuparia*), an introduced species that has naturalized well as its berries are a favourite with birds. The European ash, or rowan, is rich in history. In Britain it was planted near homes to protect owners from witches and in cemeteries to keep the dead in their graves. Christ is believed to have been crucified on a cross made from mountain ash, cedar, holly pine or cypress.

■HABITAT Native ash stays primarily where its name suggests — in the mountains. The European ash can be found at lower elevations, especially near townships.

■LOCAL SITES Good selection between Brandywine Falls and Cal-Cheak Suspension Bridge; mid to upper Blackcomb; Whistler, Cougar and Garibaldi mountains.

REDSTEM CEANOTHUS
Ceanothus sanguineus • Buckthorn family: *Rhamnaceae*

■DESCRIPTION Redstem ceanothus is a deciduous bush to 3 m in height. Its tiny fragrant flowers grow in terminal clusters 10-15 cm long. The oval leaves are finely toothed, to 10 cm long, with 3 major veins. The species name *sanguineus* means "bloody red," referring to the colour of the new twigs.

■HABITAT Open forests and edges on dry sites, at low to mid elevations.

■NATIVE USE The wood was sometimes used in smoking deer meat.

■LOCAL SITES Nairn Falls to One Mile Lake. Snowbrush (*C. velutinus*) is also in this area: It has evergreen leaves and blooms at about the same time, from the end of May to June.

Nairn Falls >

SITKA ALDER

Alnus sinuata • Birch family: *Betulaceae*

■DESCRIPTION The Sitka alder is a deciduous shrub or small tree 3-7 m high. Its coarse leaves are double-serrated, grass green and 7-10 cm long. In early spring it becomes covered in pollen-producing catkins 10-15 cm long and female cones 2 cm long.

■HABITAT As with most alders, it prefers moist conditions, from the coast of the Arctic Circle to the high mountains of California. Often grows on avalanche sites.

155

■NATIVE USE The soft straight-grained wood is easily worked and was used for making masks, bowls, rattles, paddles and spoons. The red bark was used to dye fishnets, buckskins and basket material.

■LOCAL SITES Common in the Nairn Falls area. Flowers in May.

Alnus rubra • Birch family: *Betulaceae*

■DESCRIPTION The largest native alder in North America, the red alder grows quickly and can reach 25 m in height. Its leaves are oval-shaped, grass green, 7-15 cm long, with a coarsely serrated edge. Hanging male catkins, 7-15 cm long, decorate the bare trees in early spring. The fruit (cones) are 1.5-2.5 cm long; they start off green, then turn brown and persist through winter. The bark is thin and grey on younger trees, scaly when older. Red alder leaves give a poor colour display in autumn, mainly green or brown.

■HABITAT Moist wooded areas, disturbed sites and stream banks at low to mid elevations.

■NATIVE USE The soft straight-grained wood is easily worked and was used for making masks, bowls, rattles, paddles and spoons. The red bark was used to dye fishnets, buckskins and basket material.

■LOCAL SITES Brandywine Falls, Alta and Nita lakes, Whistler Village. Larger trees can be seen around One Mile Lake. Red alder is replaced by sitka alder (*A. sinuata*; see page 155) at higher elevations.

PAPER BIRCH or CANOE BIRCH
Betula papyrifera • Birch family: *Betulaceae*

■DESCRIPTION Paper birch is a medium-size tree reaching heights of 20 m. Its serrated leaves are 8-12 cm long, rounded at the bottom and sharply pointed at the apex. Male and female catkins can be seen in early spring just before the leaves appear. The white peeling bark is a good identifier on younger trees. There is a red-bark variety that can be confused with the native bitter cherry (*Prunus emarginata*). The species name *papyrifera* means "to bear paper."

159

■HABITAT Rare in low-elevation coastal forests; common in interior forests; prefers moist soil and will tolerate wet sites.

■NATIVE USE The bark was used to make canoes, cradles, food containers, writing paper and coverings for teepees. The straight-grained wood was used for arrows, spears, snowshoes, sleds and masks.

■LOCAL SITES Large specimens can be seen at Brandywine Falls, Nairn Falls and One Mile Lake. The trail to Wedgemount Lake has very white-barked trees.

PACIFIC DOGWOOD

Cornus nuttallii • Dogwood family: *Cornaceae*

■DESCRIPTION Pacific dogwood ranges from being a multi-trunked shrub to a medium-size tree as tall as 19 m. Its leaves are elliptical, deep green above, lighter green below, to 10 cm long. The flowers are not quite as they seem: the 4–7 showy white petals are actually bracts that surround small, greenish white flowers. Clusters of small red berries 1 cm across appear by late summer. The bark is dark brown and smooth on young trees, scaly and ridged on older ones. The flower is the floral emblem of B.C., and the tree is protected by law. The painter and ornithologist John James Audubon named this tree after his friend Thomas Nuttall, the first person to classify it as a new species.

■HABITAT Coastal forests at low elevations.

■NATIVE USE The hard wood was used to make bows, arrows, harpoon shafts and, more recently, knitting needles.

■LOCAL SITES Common in the Nairn Falls area. Flowers in May.

COTTONWOOD

Populus balsamifera ssp. *trichocarpa* •
Willow family: *Salicaceae*

■DESCRIPTION Cottonwood is the tallest deciduous tree in the Pacific Northwest. It is also one of the fastest-growing, attaining heights of 45 m and trunk diameters of 2-3 m. Its leaves are heart-shaped, alternating, dark shiny green above, pale green below, 7-15 cm long. The tiny seeds on female trees hang on 7- to 13-cm catkins and are covered in white fluffy hairs known as "cotton." The deeply furrowed bark and large sticky buds are good identifiers in winter. In early summer, bits of cotton can be seen filling the skies, transporting the seeds many kilometres away from the parent trees. The wood is used commercially to make tissue paper.

■HABITAT Low moist to wet areas across B.C. and Washington. Requires sunshine and will not tolerate heavy shade.

■NATIVE USE The sticky gum from the buds was boiled and used to stick feathers on arrow shafts and to waterproof baskets and birchbark canoes.

■LOCAL SITES Common; larger trees can be seen around the parking lot and trails to Lost Lake.

Pinus contorta var. *contorta* • Pine family: *Pinaceae*

■DESCRIPTION Depending on where they are growing, shore pines vary dramatically in size and shape. By the shoreline they are usually stunted and twisted from harsh winds and nutrient-deficient soil. A little farther inland they can be straight-trunked to 20 m in height. The small cones, 3-5 cm long, are often slightly lopsided and remain on the tree for many years. The dark green needles are 4-7 cm long and grow in bundles of two. The nuts are edible but small and hard to reach. Another variety, lodgepole pine (*P. contorta* var. *latifolia*), grows straighter and taller, to 40 m. This variety is more commonly found in the Whistler area.

■HABITAT The coastal variety grows in the driest and wettest sites, from low to high elevations.

■NATIVE USE The straight wood was used for teepee poles, torches and arrow and harpoon shafts.

■LOCAL SITES Common; pure dwarf stands between Cal-Cheak Suspension Bridge and Brandywine Falls; roadsides from Brandywine Falls to Whistler Village.

WHITEBARK PINE
Pinus albicaulis • Pine family: *Pinaceae*

■DESCRIPTION Whitebark pine can reach heights of 20 m, but is often seen as a stunted bush under 3 m. Its needles are slightly curved, to 8 cm long, and grow 5 to the bundle. The cones are 5-8 cm long and purplish when young.
■HABITAT Exposed dry sites at subalpine elevations.
■LOCAL SITES Common on upper Whistler and Blackcomb mountains; several large trees (over 40 cm in diameter) can be seen at Seventh Heaven. Western white pine (*P. monticola*) can be seen at slightly lower elevations.

WESTERN WHITE PINE
Pinus monticola • Pine family: *Pinaceae*

■DESCRIPTION Western white pine is a medium-size symmetrical conifer to 40 m in height. Its bark is silvery green grey when young, dark brown and scaly when old. The cones are 15-25 cm long and slightly curved. The bluish green needles are 5-10 cm long and grow in bundles of 5. The species name *monticola* means "growing on mountains."

■HABITAT On the southern coast it grows on moist to wet soils at low to high elevations.

■NATIVE USE The bark was peeled in strips and sewn together with roots to make pine-bark canoes. The pitch was used for waterproofing.

■LOCAL SITES Lots of 18-m pines can be seen trailside to Lost Lake from Blackcomb Creek; smaller ones are around Alpha and Alta lakes, trailside to Rainbow, Wedgemount and Showh lakes.

PACIFIC SILVER FIR
Abies amabilis • Pine family: *Pinaceae*

■DESCRIPTION Pacific silver fir is a straight-trunked, symmetrical conifer to 60 m in height. The bark on young trees is smooth grey, with prominent vertical resin blisters. As the tree ages the bark becomes scaly, rougher and often lighter in colour. The cones are dark purple, barrel-shaped to 12 cm long, and sit erect on the upper portion of the tree. The needles are a lustrous dark green on the upper surface, silvery white below, with a notched tip. Pacific silver fir is one of Canada's stateliest conifers. The species name *amabilis* means "lovely fir."

■HABITAT Moist forests at mid to high elevations.

■NATIVE USE The soft wood was used for fuel, but little else. The sap was enjoyed as a chewing gum.

■LOCAL SITES Trails around Cal-Cheak Suspension Bridge; large specimens along the Cheakamus and Wedgemount lakes trails.

.DOUGLAS FIR
Pseudotsuga menziesii • Pine family: *Pinaceae*

■DESCRIPTION Douglas fir is a tall, fast-growing conifer to 90 m in height. Its bark is thick, corky and deeply furrowed. The ovate cones are 7-10 cm long and have 3 forked bracts protruding from the scales; the cones hang down from the branches, unlike true firs' cones, which stand up. The needles are 2-3 cm long, pointed at the apex, with a slight groove on the top and two white bands of stomata on the underside. The common name commemorates the botanist and explorer David Douglas.

173

■HABITAT Can tolerate dry to moist conditions from low to high elevations. Reaches its tallest size near the coast.

■NATIVE USE The wood was used for teepee poles, smoking racks, spear shafts, fishhooks and firewood.

■LOCAL SITES Large old specimens trailside to Cheakamus, Garibaldi and Wedgemount lakes; common to mid levels, where it is then taken over by Pacific silver (*Abies amabilis*; see page 171) and subalpine firs (*Abies lasiocarpa*; see page 169).

WESTERN HEMLOCK
Tsuga heterophylla • Pine family: *Pinaceae*

■DESCRIPTION Western hemlock is a fast-growing pyramidal conifer to 60 m in height. Its reddish brown bark becomes thick and deeply furrowed on mature trees. The plentiful cones are small (2-2.5 cm long), conical and reddish when young. The flat, light green leaves vary in size from 0.5 to 2 cm long. The main leaders and new shoots are nodding, giving the tree a soft, pendulous appearance that is good for identification. Western hemlock is the state tree of Washington.

■HABITAT Flourishes on the Pacific coast from Alaska to Oregon and from low levels to 1,000 m, where it is replaced by mountain hemlock (*T. mertensiana*; see page 176).

■NATIVE USE The wood has long been used for spear shafts, spoons, dishes, roasting spits and ridgepoles. The bark was boiled to make a red dye for wool and basket material.

■LOCAL SITES Large specimens trailside to Cheakamus and Garibaldi lakes.

MOUNTAIN HEMLOCK

Tsuga mertensiana • Pine family: *Pinaceae*

■**DESCRIPTION** Mountain hemlock is a smaller, stiffer tree than western hemlock (*T. heterophylla*; see page 175). It is often stunted by its harsh environment, but with good conditions it can grow slowly to 40 m in height. The reddish brown bark is rough and deeply ridged, even on young trees. The cones are up to 8 cm long, compared with 2.5 cm for the western hemlock. The bluish green leaves are equal lengths, 2-5 cm long. The species name commemorates the German botanist Franz Karl Mertens.

■**HABITAT** Typically found at higher elevations, associating with Pacific silver fir (*Abies amabilis*; see page 171) and subalpine fir (*A. lasiocarpa*; see page 169). Well-adapted to the short growing season and heavy snow packs.

■**LOCAL SITES** From the chairlift up Whistler or Blackcomb mountains a transition can be seen: midway up, western hemlock (*T. heterophylla*) is taken over by mountain hemlock.

COMMON JUNIPER

Juniperus communis • Cypress family: *Cupressaceae*

■**DESCRIPTION** Common juniper is a prostrate conifer that rarely exceeds 1 m in height and 4 m in diameter. Its bluish green needles are very sharp and prick when handled. The light green "berries" (cones) turn a dark bluish black in the second year. These mature "berries" are used to flavour gin. Common juniper is the only circumpolar conifer in the Northern Hemisphere.

■**HABITAT** Found on dry rocky outcrops from low to alpine elevations, and occasionally in bogs.

■**NATIVE USE** The wood was only used medicinally, as it rarely attains a size large enough for woodworking or carving.

■**LOCAL SITES** Harmony Lake, Musical Bumps Trail, Seventh Heaven, cliff faces across from Green Lake.

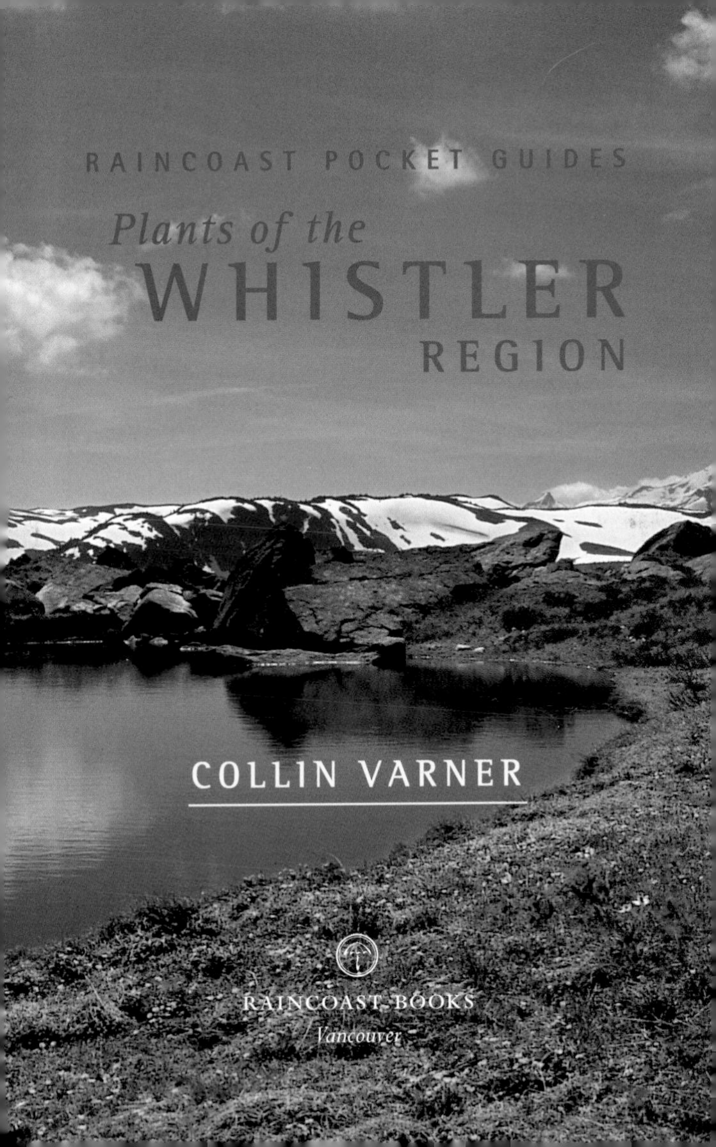

RAINCOAST POCKET GUIDES

Plants of the
WHISTLER
REGION

COLLIN VARNER

RAINCOAST BOOKS
Vancouver

Raincoast Books acknowledges the ongoing financial support of the
Government of Canada through The Canada Council for the Arts and
the Book Publishing Industry Development Program (BPIDP); and the
Government of British Columbia through the BC Arts Council.

Edited by Simone Doust
Series design by Ingrid Paulson
Layout by Teresa Bubela

National Library of Canada Cataloguing in Publication Data
Varner, Collin
  Plants of the Whistler region / Collin Varner.

  (Raincoast pocket guides)
  Includes bibliographical references and index.
  ISBN 1-55192-602-4

  1. Botany—British Columbia—Whistler Region. I. Title. II. Series.
QK203.B7V372 2003   581.9711'31   C2002-911415-2

Library of Congress Catalogue Number: 2002096043

Raincoast Books   *In the United States:*
9050 Shaughnessy Street   Publishers Group West
Vancouver, British Columbia   1700 Fourth Street
Canada   V6P 6E5   Berkeley, California
www.raincoast.com   94710

Printed in Hong Kong, China by Book Art Inc., Toronto

10 9 8 7 6 5 4 3 2 1

TABLE OF CONTENTS

Dressed in Green 7

Flowers 11

Berries 91

Ferns 106

Rogues 114

Shrubs & Bushes 131

Broadleaf Trees 152

Conifers 164

Glossary/Plant Parts, Leaf Shapes 183

Bibliography 185

Index 186

To most people, the word "Whistler" conjures up thoughts of wonderful snowfalls and winter sports, but Whistler is quickly expanding as an all-season destination with endless opportunities for outdoor recreation such as hiking, cycling, canoeing and walking. This book is a handy guide to Whistler and the surrounding region when it is dressed in green.

The region has the advantage of a broad growing range from low to alpine elevations — a combination that makes it one of North America's most diversified growing areas. This range of elevation also means that the same plants flower at different times throughout the region. For every 305 metres of rise, for example, there is at least a week's delay in flowering time, so red elderberry flowers in mid-April at low levels, but not until mid-July on Blackcomb Mountain.

The lower-elevation areas are to the southwest of Whistler at Brandywine Falls and to the northeast at Nairn Falls. In both of these areas, the warmer temperatures allow for lush growth and giant trees. The first plants to flower are wild ginger, devil's club, wintergreen, orchid, honeysuckle, trailing violet, sarsaparilla and tiger lily. The plants at mid levels from Whistler Village to lower Singing Pass emerge later, while the subalpine to alpine meadows and slopes wait until July, August and even September to flower. The short growing season at the upper

< *Whistler Mountain*

elevations forces the plants to flower simultaneously, with the best viewing time falling between mid-July and mid-August.

This book describes more than 150 of the plants most likely to be encountered when walking or hiking in the Whistler-Blackcomb region. Each species has a fact sheet with three or four entries. DESCRIPTION presents the plant and how to identify it, while HABITAT explains where it grows best. Plants valued by First Nations peoples have an entry for NATIVE USE. In the LOCAL SITES section, I have listed a few places where each

plant can be found and its flowering time. Most of the place names in the LOCAL SITES entries can be found in the maps on the front and back flaps.

Please note that I have only included native plants and introduced species that survive and thrive in the wild. Also note that to keep the book pocket-sized, I have not included every known species and have been selective with the information given for each cited plant. The observer with a keen interest and a sharp eye can expect to discover more treasures.

— Collin Varner

Harmony Lake

ACKNOWLEDGEMENTS

This guide could not have been completed without the help of many people. I would like to acknowledge and thank the following:

My hiking companions Patrick Murphy, Cathy Sinclair, Allan Davidson, Richard Davidson and my wife, Wendy.

The administration at Whistler-Blackcomb, with special thanks to Kim Muller.

Bob Brett, director of the Whistler Naturalists Society, for his keen plant identification and proofreading.

Paul Beswetherick, Landscaping Supervisor for the Municipality of Whistler, for his skillful proofreading.

Brenda O'Reilly, for her superb keyboarding skills.

The University of British Columbia, for its continuing support.

Savio Otis, owner of Whistler Cross Country Ski & Hike.

The great staff at Raincoast Books: Scott Steedman, Simone Doust, Ingrid Paulson, Teresa Bubela and Marjolein Visser.

TWINFLOWER
Linnaea borealis • Honeysuckle family: *Caprifoliaceae*

■ DESCRIPTION Twinflower is an attractive trailing evergreen to 10 cm in height. Its nodding pink flowers are fragrant, to 5 mm long, and borne in pairs at the end of slender, Y-shaped stems. The evergreen leaves are 1 cm long, oval, shiny dark green above and paler below, with minute teeth on the upper half. The genus *Linnaea* is named for Carolus Linnaeus, Swedish botanist and founder of the binomial system for plant and animal classification. Twinflower is said to have been his favourite flower.

■ HABITAT Common in low to mid elevation forests across Canada.

■ LOCAL SITES Common; carpets are found around the observation deck at Brandywine Falls, at mid levels in Singing Pass and draped over the moss-covered rocks along Nita and Alta lakes. Flowers start mid-June at lower elevations and continue into August at higher elevations.

< *View from top of Wedgemount Lake Trail*

11

CLIMBING HONEYSUCKLE
Lonicera ciliosa • Honeysuckle family: *Caprifoliaceae*

■ DESCRIPTION Climbing honeysuckle is a deciduous woody vine capable of climbing trees to 8 m in height. Its orange flowers are trumpet-shaped, to 4 cm long, and form in clusters in the terminal leaves. By late summer bunches of bright red berries are produced in the cup-shaped leaves. The leaves are oval, 5-8 cm long and, like all honeysuckles, opposite. This species is the showiest of the native honeysuckles. Its main pollinators are hummingbirds and moths. CAUTION: the berries are considered poisonous.

■ HABITAT Scattered in low-elevation Douglas fir forests, more common near the ocean and in the Gulf and San Juan islands.

■ NATIVE USE The vines were used to weave mats, blankets and bags.

■ LOCAL SITES Common around Nairn Falls. Flowers from the end of May to June.

CANADA THISTLE
Cirsium arvense • Aster family: *Asteraceae*

■ DESCRIPTION Canada thistle is an armed herbaceous perennial from 0.7 to 1.5 m in height. Its flowers are more abundant than the bull thistle's (*C. vulgare*) but are smaller (1 cm across) and a lighter shade of purple. Canada thistle is dioecious, meaning the male and female flowers grow on separate plants. The irregularly shaped leaves are green on top, white and hairy beneath and have spiny edges; this thistle does not have spiny wings on its stems. Despite its name, Canada thistle is a native of Europe. The species name *arvense* means "of cultivated fields."

■ HABITAT Common on wasteland and in cultivated areas, at low to mid levels.

■ LOCAL SITES Seen on roadsides throughout Whistler Village and on the lower chairlift slopes. Flowers with bull thistle from the end of July through August.

13

BULL THISTLE
Cirsium vulgare • Aster family: *Asteraceae*

■ DESCRIPTION Bull thistle is an introduced, well-armed biennial to 1.7 m in height. Its showy flowers are pinkish purple, well-armed towards the base, to 4 cm long. The leaves are alternate, deeply incised and armed on both the edges and the top surface. This species puts out vegetative growth in the first year and flowers in the second. The species name *vulgare* means "common."

■ HABITAT Disturbed sites. Because grazing animals do not eat these plants, they have spread well into fertile pastures and fields.

■ LOCAL SITES Common along the Sea to Sky Highway and in Whistler Village. Flowers from the end of July through August.

14

WALL LETTUCE

Lactuca muralis • Aster family: *Asteraceae*

■ **DESCRIPTION** Wall lettuce is an introduced herbaceous biennial to 1.5 m tall. Its tiny yellow flowers, which resemble small dandelions, grow in loose clusters to 25 cm across. The fruit (achenes) are small and covered with fluffy hairs. Leaves are variable in size and shape, though most are deeply incised and clasp the stem. The milky sap gave the plant its genus name, *Lactuca,* from the Latin word for milk, *lac.*

■ **HABITAT** Very common in southern B.C., on roadsides and highway medians and in open forests.

■ **LOCAL SITES** Roadsides from Brandywine Falls to Nairn Falls. Also common at forest edges around Whistler Village. Flowers July through September.

OXEYE DAISY

Chrysanthemum leucanthemum • Aster family: *Asteraceae*

■ **DESCRIPTION** Oxeye daisy is an aromatic herbaceous perennial to 75 cm tall. Its flowers have the typical daisy white ray petals and yellow centre disks, to 5 cm across. The basal leaves are obovate with rounded teeth; the stem leaves are similar, though alternate. This is a European introduction that has naturalized in most of the Pacific Northwest. "Chrysanthemum" is from the Greek *chrysos* ("gold") and *anthos* ("flower").

■ **HABITAT** Fields, meadows, very common on roadsides.

■ **LOCAL SITES** Part of the roadside vegetation in the Whistler region. Flowers abundantly from mid-June to August.

16

SWEET COLTSFOOT or ALPINE COLTSFOOT
Petasites frigidus • Aster family: *Astercaceae*

■ DESCRIPTION Sweet coltsfoot is a herbaceous perennial to 50 cm in height. Its white to pink flowers are grouped together to form open clusters held above the leaves, which are white-woolly below and deeply lobed into 3-5 segments. The genus name *Petasites* comes from the Greek word *petasos*, meaning "hat," referring to the large basal leaves. Japanese children once used the large leaves of *Petasites japonica* as hats.

17

■ HABITAT Bogs, lake edges, wet meadows at subalpine to alpine elevations.

■ NATIVE USE The leaves were used to cover berries in steam cooking pits.

■ LOCAL SITES Common in the Harmony Lake area and wet meadows in upper Singing Pass. Flowers late July to August.

ARROW-LEAFED GROUNDSEL

Senecio triangularis • Aster family: *Asteraceae*

■ **DESCRIPTION** Arrow-leafed groundsel is a clump-forming herbaceous perennial 30-150 cm in height. Its yellow flowers grow to 1.5 cm across and form in flat-topped clusters. The triangular leaves are 4-20 cm long, squared at the base and evenly toothed.

■ **HABITAT** Moist sites from mid to alpine elevations.

■ **LOCAL SITES** Common; found in Singing Pass and Harmony Bowl. Flowers July, August and September, depending on elevation.

18

YARROW

Achillea millefolium • Aster family: *Asteraceae*

■ DESCRIPTION Yarrow is a herbaceous perennial to 1 m in height. Its many small white flowers form flat-topped clusters 5-10 cm across. The aromatic leaves are so finely dissected that they appear fern-like, hence its species name "a thousand leaves." The genus is named after Achilles, a hero of Greek mythology.

■ HABITAT Roadsides, wasteland, common at low to mid elevations.

■ NATIVE USE Infusions and poultices were made for cold remedies.

■ LOCAL SITES Widespread from the summit of Whistler Mountain (Harmony Bowl) to lower elevations around Alpha and Green lakes, with a great display of flowers above Nairn Falls in mid-June. Flowers July to August at higher elevations.

CANADA GOLDENROD

Solidago canadensis • Aster family: *Asteraceae*

■ **DESCRIPTION** Canada goldenrod is a herbaceous perennial of various heights, from 30 to 150 cm. Its small golden flowers are densely packed to form terminal pyramidal clusters. The many small leaves grow at the base of the flowers; they are alternate, lance-linear, sharply saw-toothed to smooth.

■ **HABITAT** Roadsides, wasteland, forest edges at low to mid elevations.

20

■ **LOCAL SITES** Found along the Sea to Sky Highway roadsides from Lions Bay to Whistler's lower lakes. Flowers July and August.

PEARLY EVERLASTING

Anaphalis margaritacea • Aster family: *Asteraceae*

■ **DESCRIPTION** Pearly everlasting grows to 80 cm in height and produces heads of small yellowish flowers surrounded by dry white bracts. The leaves are lance-shaped, green above and covered with a white felt underneath. If picked before they go to seed, the flowers remain fresh-looking long after they are brought in.

■ **HABITAT** Common on disturbed sites, roadsides and rock outcrops.

■ **LOCAL SITES** Common on exposed sites around the Showh Lakes area, Musical Bumps Trail to Singing Pass and along roadsides and cliff faces in the Whistler region. Flowers begin mid-July at mid elevations.

SUBALPINE DAISY or MOUNTAIN DAISY
Erigeron peregrinus • Aster family: *Asteraceae*

■ **DESCRIPTION** Mountain daisy is a showy herbaceous perennial to 60 cm in height. Its lavender to purplish flowers grow to 6 cm across and are usually borne singularly. The lance-shaped basal leaves vary from 1 to 20 cm long and form in clumps; smaller lance-shaped leaves clasp the flower stem. This beautiful plant is often seen in association with arctic lupine (*Lupinus arcticus*; see page 79), valerian (*Valeriana sitchensis*; see page 77) and arnica (*Arnica latifolia*; see page 23).

22

■ **HABITAT** Moist meadows or slopes at mid to subalpine elevations.

■ **LOCAL SITES** Common along Musical Bumps Trail and at Harmony Bowl, Symphony Bowl, Piccolo Summit and Seventh Heaven. Flowers mid-July through August and into September at higher elevations.

MOUNTAIN ARNICA

Arnica latifolia • Aster family: *Asteraceae*

■ DESCRIPTION Mountain arnica is an unbranched herbaceous perennial to 60 cm in height. Its showy yellow flowers grow to 5 cm across and light up the mountain slopes by midsummer; they usually grow in groups of 3, each with 8-12 ray petals. The stem leaves are coarsely toothed, broadly lanceolate and form in 2-4 opposite pairs. Mountain arnica is usually seen with 1 terminal flower and 2-3 more on the lower secondary stems.

■ HABITAT One of the most common arnicas; prefers moist slopes and meadows at mid to subalpine elevations.

■ LOCAL SITES Common on the upper slopes of Whistler and Blackcomb mountains. Flowers July and August.

23

ALPINE PUSSYTOES
Antennaria alpina • Aster family: *Asteraceae*

■ DESCRIPTION Alpine pussytoes is a mat-forming herbaceous perennial to 12 cm in height. Its flowers are creamy white and held in tight terminal clusters. The leaves are soft, woolly white and mainly basal. As the common name implies, the flower heads resemble cat paws.

■ HABITAT Dry, exposed sites at subalpine to alpine elevations.

■ LOCAL SITES Common on rocky slopes around Harmony Bowl. Flowers from the end of July to August.

YELLOW SALSIFY
Tragopogon dubius • Aster family: *Asteraceae*

■ DESCRIPTION Yellow salsify is an introduced biennial to 1 m in height. Its lemon yellow flowers are 4-5 cm across, with bright green bracts protruding beyond the rays; by July and August the flowers turn into large dandelion-like seed heads to 7 cm across. The long grass-like leaves clasp the stems at the base. The genus name comes from the Greek *tragos*, meaning "goat," and *pogon*, meaning "beard," referring to the appearance of the silky seeds.

■ HABITAT Dry roadsides, disturbed sites and meadows.
■ LOCAL SITES Scattered on roadsides from Green Lake to Nairn Falls. A wonderful display of hundreds of flowers can be seen above Nairn Falls from June to July.

ORANGE HAWKWEED or DEVIL'S PAINTBRUSH
Hieracium aurantiacum • Aster family: *Asteraceae*

■ **DESCRIPTION** Orange hawkweed is an introduced herbaceous perennial to 60 cm in height. Its flowers are a dazzling orange-red and grow to 2.5 cm across. The leaves are lanceolate, mainly basal and, like the stem, slightly hairy. This plant's invasive rhizomes and winged seeds have allowed it to spread rapidly in southern B.C. The species name *aurantiacum* means "orange-red."

■ **HABITAT** Dry roadsides, wasteland, disturbed sites from low to high elevations.

■ **LOCAL SITES** Tens of thousands electrify the sides of the Sea to Sky Highway between Squamish and Pemberton in mid- to late June. Also found on roadsides approaching the Garibaldi Lake trailhead and on One Mile Lake Trail.

WHITE-FLOWERED HAWKWEED

Hieracium albiflorum • Aster family: *Asteraceae*

■ DESCRIPTION White-flowered hawkweed is a herbaceous perennial from 40 cm to 1 m tall. Its creamy white flowers are 1 cm across and, like the dandelion's, the head is made up of only ray flowers. The leaves are mainly basal, lanceolate and hairy. The name "hawkweed" comes from a Greek myth that the juice of the plant would clear the eyes of a hawk.

■ HABITAT Disturbed sites, open coniferous forests at low to mid elevations.

■ LOCAL SITES Dry, open forests above Brandywine Falls, lower trailsides to Rainbow Falls and Whistler Village.

PATHFINDER

Adenocaulon bicolor • Aster family: *Asteraceae*

■ **DESCRIPTION** Pathfinder is a herbaceous perennial to 1 m tall. Its tiny white flowers are inconspicuous compared with the large (10-15 cm long) bicoloured, heart-shaped leaves. The mature seeds are hooked, allowing them to attach to passing animals and people's clothing. The leaves flip over when walked through, revealing the silvery underside and thus marking the path.

■ **HABITAT** Shaded forests at low to mid elevations.

■ **LOCAL SITES** Brandywine Falls, Cheakamus Lake area, trailside to Rainbow Falls and One Mile Lake. Flowers June to July.

YELLOW CYPRESS or YELLOW CEDAR

Chamaecyparis nootkatensis • Cypress family: *Cupressaceae*

■**DESCRIPTION** Yellow cypress is a large, slow-growing conifer of conical habit that often exceeds 45 m in height. Its thin, greyish brown bark can be shed in vertical strips. The reddish brown cones are round, to 1-2 cm across, with 4-6 scales tipped with pointed bosses (red cedar has egg-shaped cones). The bluish green leaves are prickly to touch and more pendulous than the red cedar's. Yellow cypress was first documented in Nootka Sound on the west coast of Vancouver Island in 1791 by Archibald Menzies, hence the species name. Its genus name is now being changed to *Cupressus*.

■**HABITAT** On the southern coast it grows in moist forests at mid to high elevations.

■**NATIVE USE** The wood was used for carving fine objects such as bentwood boxes, chests and intricately carved canoe paddles. The bark is softer than red cedar's and women used it to make clothing, blankets, baskets, rope and hats.

■**LOCAL SITES** Common at low to high elevations on Whistler and Cougar mountains; Cal-Cheak Suspension Bridge has trees reaching 1.5 m in diameter; upper trails to Rainbow and Wedgemount lakes.

< *The distinct pendulous profile of the yellow cypress, on the right in this photo (top)*

SITKA SPRUCE
Picea sitchensis • Pine family: *Pinaceae*

■DESCRIPTION Sitka spruce is often seen on rocky outcrops as a twisted dwarf tree, though in favourable conditions it can exceed 90 m in height. Its reddish brown bark is thin and patchy, a good identifier when the branches are too high to observe. The cones are gold brown, to 8 cm long. The needles are dark green, to 3 cm long and sharp to touch. Sitka spruce has the highest strength-to-weight ratio of any B.C. or Washington tree. It was used to build the frame of Howard Hughes' infamous plane *Spruce Goose.*

■HABITAT A temperate rainforest tree that does not grow farther than 200 km from the ocean.

■NATIVE USE The new shoots and inner bark were a good source of vitamin C. The best baskets and hats were woven from spruce roots, and the pitch (sap) was often chewed as a gum.

■LOCAL SITES Sitka spruce has hybridized with Engelmann spruce (*P. engelmannii*) in a narrow strip in the Whistler region. Locals refer to it as the "Whistler hybrid" or "Whistler spruce."

WESTERN RED CEDAR
Thuja plicata • Cypress family: *Cupressaceae*

■DESCRIPTION Western red cedar is a large, fast-growing conifer reaching over 60 m in height. Its bark sheds vertically and ranges from cinnamon red on young trees to grey on mature ones. The bases of older trees are usually heavily flared, with deep furrows. The egg-shaped cones are 1 cm long, green when young, turning brown and upright when mature. The bright green leaves are scale-like, with an overlapping-shingle appearance. Western red cedar is B.C.'s provincial tree.

■HABITAT Thrives on moist ground at low elevations. Tolerates drier or higher sites but doesn't grow as big.

■LOCAL SITES Common at low to mid elevations. Huge specimens found in Ancient Cedars Grove on Cougar Mountain.

Anther	The pollen-bearing (top) portion of the stamen.
Axil	The angle made between a stalk and a stem on which it is growing.
Biennial	Completing its life cycle in two growing seasons.
Boss	Knob-like studs, as in the points on cones of yellow cypress.
Bract	A modified leaf below the flower.
Calyx	The collective term for sepals, the outer part of the flower.
Catkin	A spike-like or drooping flower cluster, male or female (e.g., cottonwood).
Corm	An underground swollen stem capable of producing roots, leaves and flowers.
Deciduous	A plant that sheds its leaves annually, usually in the autumn.
Dioecious	Male and female flowers on separate plants.
Epiphyte	A plant that grows on another plant for physical support, without robbing the host of nutrients.
Herbaceous perennial	A nonwoody plant that dies back to the ground each year and regrows the following season.
Lenticel	Raised organs that replace stomata on a stem.
Node	The place on a stem where the leaves and auxiliary buds are attached.
Obovate	Oval in shape, with the narrower end pointing downward, like an upside down egg.
Panicle	A branched inflorescence.
Petiole	The stalk of a leaf.
Pinnate	A compound leaf with the leaflets arranged on both sides of a central axis.
Pinnule	Leaflet of a pinnately compound leaf.
Rhizome	An underground modified stem. Runners and stolons are on top of the ground.

Abies
  amabilis, 171, 173, 176
  lasiocarpa, 168, 173, 176
Acer
  circinatum, 154
  glabrum, 154
  macrophyllum, 153
Achillea millefolium, 19
Actaea rubra, 60
Adenocaulon bicolor, 28
Adiantum pedatum, 107
alder
  red, 157
  sitka, 155, 157
Alnus
  rubra, 157
  sinuata, 155, 157
Alpha Lake, 19, 39, 70, 96, 101, 103, 115, 117, 127, 133, 167, 168
Alta Lake, 11, 37, 91, 96, 101, 103, 136, 138, 140, 148, 157, 167
alumroot
  small-flowered, 42
  smooth, 42
Amelanchier alnifolia, 96
Ancient Cedars Grove, 74, 145, 182
Ancient Cedars Trail, 31
Anemone occidentalis, 59, 79
anemone, western, 59, 79
Antennaria alpina, 24
Apocynum androsaemifolium, 81
Aquilegia formosa, 57
Aralia nudicaulis, 125
Arctostaphylos uva-ursi, 91
Arnica latifolia, 22, 23
arnica, mountain, 22, 23
Aruncus dioicus, 90
Asarum candatum, 130
Athyrium filix-femina, 69, 108, 111
avens, large-leafed, 38
azalea, false, 138

baneberry, 60
bearberry, 91
Betula papyrifera, 159
birch
  canoe, 159
  paper,159

blackberry,
  cutleaf, 99
  Himalayan , 99
  trailing, 99
blackcap, 101
Blackcomb Creek, 141, 167
Blackcomb Lake, 56, 62
Blackcomb Mountain, 23, 31, 40, 41, 47, 53, 66, 87, 88, 89, 105, 115, 130, 135, 139, 141, 149, 166, 169, 176
Blechnum spicant, 109
bleeding heart, Pacific, 63
blueberry,
  Alaskan, 95
  oval-leafed, 95
Body Bag Bowl, 50
bog laurel, western, 139, 140
bog orchid, white, 88
bogbean, 122
bramble, five-leafed, 104
Brandywine Falls, 11, 15, 27, 28, 30, 35, 36, 39, 43, 46, 54, 55, 70, 75, 83, 86, 91, 92, 96, 97, 99, 100, 102, 104, 105, 112, 116, 118, 119, 120, 122, 123, 127, 128, 130, 131, 132, 134, 135, 136, 138, 139, 140, 145, 146, 153, 157, 159, 165, 168
buckbean, 122
bunchberry, 67
buttercup
  mountain, 56
  subalpine, 56

Cal-Cheak Suspension Bridge, 110, 118, 120, 122, 127, 128, 131, 140, 149, 165, 171, 179
Caltha leptosepala, 62
Calypso bublosa, 84
camas
  common, 34
  death, 34
Camassia quamash, 34
Canada goldenrod, 20
Cassiope mertensiana, 142
Castilleja sp., 47
cat-tail, 117
Ceanothus
  Sanguineus, 150

Velutinus, 150
ceanothus, redstem, 150
cedar
 ground, 119
 western red, 182
 yellow, 179
Chamaecyparis nootkatensis, 178
Cheakamus Lake, 28, 33, 36, 37, 42, 43,
 44, 53, 57, 60, 61, 63, 67, 70, 72, 82,
 87, 93, 121, 130, 145, 154, 171, 173,
 175
Cheakamus Lake Trail, 75, 101
Cheakamus River, 137
cherry, bitter, 159
Chimaphila
 Menziesii, 76
 Umbellata, 75, 76
Chrysanthemum leucanthemum, 16
cinquefoil, fan-leafed, 40, 56
Cirsium
 arvense, 13
 vulgare, 13, 14
Claytonia
 perfoliata, 82
 sibirica, 82
Clintonia uniflora, 30
clubmoss, running, 118
coltsfoot
 alpine, 17
 sweet, 17
columbine, red, 57
Cornus
 canadensis, 67
 nuttallii, 161
 stolonifera, 146
Corydalis sempervirens, 64
corydalis, pink, 64
cotton grass, narrow-leafed, 123
cottonwood, 163
Cougar Mountain, 31, 46, 61, 74, 137,
 138, 145, 147, 149, 154, 179, 182
cow-parsnip, 53
 giant, 53
cranberry, highbush, 105
Cryptogramma crispa, 110
currant, red-flowering, 144
cypress, yellow, 179

daisy
 mountain, 22
 oxeye, 16
 subalpine , 22
Decker Creek, 56
devil's club, 36, 145
Dicentra formosa, 63
Disporum
 hookeri, 37
 Smithii, 37
dogbane, spreading, 81
dogwood
 dwarf, 67
 Pacific, 161
dogwood, red-osier, 146
Drosera
 anglica, 116
 rotundifolia, 116
Dryopteris expansa, 107, 111

elderberry, red, 135
elder, red-berried, 135
Epilobium
 angustifolium, 65, 123
 latifolium, 66
Equisetum arvense, 115
Erigeron perigrinus, 22
Eriophorum angustifolium, 123

fairybells
 Hooker's, 37
 Smith's, 37
fairyslipper, 84
falsebox, 147
fern
 deer, 109
 lady, 69, 108, 111
 licorice, 109, 153
 maidenhair, 107
 mountain, 110
 oak, 113
 parsley, 110
 shield, 111
 spiny wood, 107, 111
 western sword, 109, 112
fir
 Douglas, 173
 pacific silver, 171, 173, 176

subalpine, 169, 173, 176
fireweed, 65
    alpine, 66
    broad-leafed, 66
Flute Summit, 80,
foam flower, one-leafed, 44
Fragaria
    vesca, 102
    virginiana, 102
fringecup, 43

Garibaldi Lake, 30, 33, 37, 42, 44, 60, 63,
    67, 71, 82, 83, 84, 85, 93, 104, 107,
    108, 111, 118, 145, 147, 154, 173, 175
Garibaldi Lake Trail, 26, 73, 75, 112, 168
Garibaldi Mountain, 141, 149
Gaultheria
    ovatifolium, 92
    shallon, 92
Geum macrophyllum, 38
ginger, wild, 130
goat's beard, 90
Goodyera oblongifolia, 83
gooseberry,
    black, 103
    black swamp, 103
Green Lake, 19, 51, 177
groundsel, arrow-leafed, 18
Gymnocarpium dryopteris, 113

hardhack, 133
Harmony Bowl, 18, 19, 22, 24, 31, 45,
    76
Harmony Lake, 17, 41, 143, 177
Harmony Meadows, 129
hawkweed
    orange, 26
    white-flowered, 27
heal-all, 55
hemlock
    western, 175, 176
    mountain, 175, 176
Heracleum
    lanatum, 53
    mantegazzianum, 53
Heuchera
    micrantha, 42
    glabra, 42

Hieracium
    aurantiacum, 26
    albiflorum, 27
honeysuckle
    climbing, 12
    Utah, 137
horsetail, common, 115
huckleberry
    black, 95
    fool's, 138
Hypopitys monotropa, 128

Indian celery, 53
Indian hellebore, 31

Jordan Creek, 70
juniper, common, 177
Juniperus communis, 177

Kalmia microphylla ssp. occidentalis,
    139, 140
kinnikinnick, 91

Lactuca muralis, 15
Ledum Groenlandicum, 139, 140
Leptarrhena pyrolifolia, 129
Lilium columbianum, 32
lily
    bead, 30
    corn, 31
    tiger, 32
Linnaea borealis, 11
Lions Bay, 20
Listera cordata, 85
Lonicera
    ciliosa, 12
    involucrata, 136
    utahensis, 137
Lost Lake, 35, 91, 117, 130, 148, 163, 167
Lotus corniculatus, 22, 78
lousewort, bracted, 52, 79
Luetkea pectinata, 41, 79
lupine, arctic, 22, 79
Lupinus arcticus, 79
Lycopodium
    clavatum, 118
    complanatum, 119
    dendroideum, 120

Lysichiton americanum, 69, 107

Mahonia
    aquifolium, 93
    nervosa, 93
maple
    bigleaf, 153
    Douglas, 154
    vine, 154
meadowrue, western, 61
Menyanthes trifoliata, 122
menzies' pipsissewa, 76
Menziesia ferruginea, 138
Mimulus
    lewisii, 49
    tilingii, 48
miner's lettuce, Siberian, 82
monkey-flower
    alpine, 48
    Lewis', 48
    mountain, 48
    pink, 49
    yellow, 48
mooseberry, 105
mountain ash
    European, 149
    sitka, 149
mountain heather
    pink, 141, 142
    white, 142
    yellow, 141
mountain lover, 147
sweet cicely, mountain, 54
Musical Bumps, 47, 56
Musical Bumps Trail, 21, 22, 41, 51, 80, 177
Myrica gale, 148

Nairn Falls, 12, 15, 19, 25, 30, 33, 34, 42, 46, 50, 51, 54, 55, 64, 71, 72, 73, 76, 81, 83, 84, 90, 93, 101, 110, 125, 130, 132, 133, 134, 150, 153, 154, 155, 159, 161
ninebark, 134
Nita Lake, 11, 37, 42, 70, 90, 103, 127, 136, 137, 138, 140, 148, 157, 168
Nuphar polysepalum, 126

Oboe Summit, 31
One Mile Lake, 28, 47, 55, 64, 71, 76, 81, 125, 133, 134, 136, 148, 150, 157, 159
One Mile Lake Trail, 26
oplopanax horridus, 36, 133, 145
Oregon grape, 93
    tall, 93
Orthilia secunda, 74
Osmorhiza chilensis, 54

Pachistima myrsinites, 147
paintbrush, 47
    devil's, 26
partridgefoot, 41, 79
pasque flower, western, 59
pathfinder, 28
pearly everlasting, 21
Pedicularis bracteosa, 52, 79
Pemberton, 26
penstemon
    davidson's , 51
    slender blue, 50
Penstemon
    davidsonii, 51
    procerus, 50
Petasites frigidus, 17
Phlox diffusa, 80
phlox, spreading, 80
Phyllodoce
    empetriformis, 141, 142
    granduliflora, 141
Physocarpus capitatus, 134
Piccolo Summit, 22, 41, 45, 48, 66, 80
Picea
    Engelmannii, 181
    sitchensis, 181
piggy-back plant, 43
pine
    ground, 120
    lodgepole, 165
    shore, 165
    whitebark, 166
    western white, 166, 167
    pinesap, 128
Pinus
    albicaulis, 166
    contorta, 165
    contorta var. latifolia, 165

monticola, 166, 167
Plantago
  lanceolata, 124
  major, 124
plantain
  common, 124
  rattlesnake, 83
  ribwort, 124
Platanthera
  dilatata, 88
  orbiculata, 86
  stricta, 87
Polypodium glycrrhiza, 109
Polystichum munitum, 109, 112
pond lily, yellow, 127
Populus balsamifera spp. trichocarp, 163
Potentilla flabellifolia, 40, 56
prince's pine, 75, 76
  little, 76
princess pine, 120
Prunella vulgaris, 55
Prunus emarginata, 159
Pseudotsuga menziesii, 173
pussytoes, alpine, 24
Pyrola
  asarifolia, 72
  chlorantha, 73
  picta, 71

queen's cup, 30

Rainbow Falls, 27, 28, 39, 55, 76, 85
Rainbow Lake, 29, 31, 36, 41, 47, 50, 62,
  63, 67, 69, 72, 75, 77, 87, 104, 105,
  107, 111, 115, 118, 129, 139, 143, 145,
  167, 179
Ranunculus eschscholtzii, 56
raspberry
  black, 101
  creeping, 104
rein-orchid
  round-leafed, 86
  slender, 87
  white, 88
Rhododendron albiflorum, 79, 143
rhododendron, white-flowered, 79, 143
Ribes
  lacustre, 103

sanguineum, 144
Rosa
  gymnocarpa, 131
  nutkana, 131
rose
  baldhip, 132
  Nootka, 131
  woodland, 132
Roundhouse, 143
Rubus
  discolor, 99
  laciniatus, 99
  leucodermis, 101
  parviflorus, 97
  pedatus, 104
  spectabilis, 100
  ursinus, 99
Russet Lake, 113

salal, 92
salmonberry, 100
salsify, yellow, 25
Sambucus
  caerlea, 135
  racemosa, 135
sarsparilla, wild, 125
Saskatoon berry, 96
Saxifraga
  ferruginea, 46, 138
  tolmiei, 45
saxifrage
  Alaska, 46
  Leatherleaf, 129
  Tolmie's, 45
Sea to Sky Highway, 14, 20, 26, 53, 81,
  88, 90, 133, 146
Sedum divergens, 89
Self-heal, 55
Senecio triangularis, 18
serviceberry, 96
Seventh Heaven, 22, 31, 40, 41, 47, 50,
  52, 56, 66, 77, 143, 166, 177
Showh Lakes, 21, 118, 137, 154, 167
Singing Pass, 11, 17, 18, 21, 29, 31, 33, 35,
  36, 43, 47, 49, 52, 57, 61, 87, 97, 99,
  104, 105, 110, 115, 136, 138, 143, 147
Singing Pass Trail, 113
skunk cabbage, 69, 107

Smilacina
  racemosa, 35, 36
  stellata, 36
snowbrush, 150
Solidago canadensis, 20
Solomon's seal
  false, 35, 36
  star-flowered, 36
Sorbus
  aucuparia, 149
  sitchensis, 149
spruce
  Engelmann, 181
  sitka, 181
Squamish, 26, 153
steeplebush, 133
stinging nettle, 121
stonecrop, spreading, 89
strawberry
  blue-leaf, 102
  wild, 102
  woodland, 102
Streptopus amplexifolius, 29
sundew
  narrow-leafed, 116
  round-leafed, 116
sweet gale, 148
Symphony Bowl, 22, 48

Taxus brevifolia, 168
tea, Labrador, 139, 140
tea-berry, western, 92
Tellima grandiflora, 43
Thalictrum occidentale, 61
thimbleberry, 97
thistle
  bull, 13, 14
  Canada, 13
Thuja plicata, 182
Tiarella trifoliata var. unifoliata, 44
Tragopogon dubius, 25
trefoil, bird's-foot, 78
Tsuga
  heterophylla, 175, 176
  mertensiana, 175, 176
twayblade, heart-leafed, 85
twinberry,
  red, 137

  black, 136
twinflower, 11
twisted stalk, 29
Typha latifolia, 117

Urtica dioica, 121

Vaccinium
  alaskaense, 95
  membranaceum, 95
  ovalifolium, 95
valerian, sitka, 22, 77
Valeriana sitchensis, 22, 77
Veratrum viride, 32
Viburnum edule, 105
Viola glabella, 70
violet, stream, 70

wall lettuce, 15
Wedgemount Lake, 22, 51, 67, 104, 111,
  113, 118, 138, 167, 171, 173, 179
Wedgemount Lake Trail, 159
Whistler Mountain, 19, 23, 31, 45, 89,
  141, 143, 147, 149, 166, 176, 179
Whistler Village, 13, 14, 15, 27, 35, 39, 88,
  91, 105, 112, 144, 146, 157, 165, 169
white marsh-marigold, 62
wintergreen
  green-flowered, 73
  one-sided, 74
  painted, 71
  pink, 72, 73
  white-veined, 71
Wizard chairlift, 88
wood betony, 52, 79

yarrow, 19
yew
  Pacific, 168
  western, 168

Zygadenus venenosus, 34

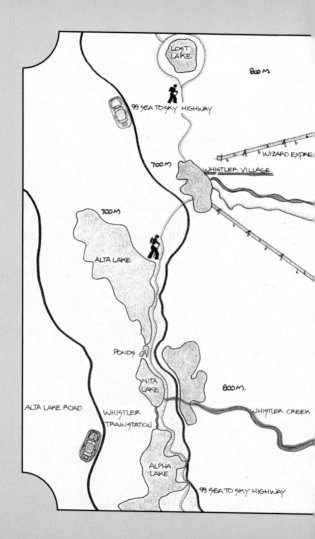

TWISTED STALK
Streptopus amplexifolius • Lily family: *Liliaceae*

■ DESCRIPTION Twisted stalk is a branching, herbaceous perennial, 1-2 m in height. Its greenish white flowers are 1 cm long and borne in leaf axils on slender twisted stalks. The fruit develops into a bright red oval berry to 1 cm long. The ovate leaves are alternate, 5-12 cm long. They clasp the stem directly, with no petiole; the species name *amplexifolius* means "clasping leaves." The flowers and fruit hang from the leaf axils along the branches.
CAUTION: the berries are considered poisonous.

■ HABITAT Cool moist forests at low to high elevations.
■ NATIVE USE The plants were tied to the clothing or hair for their scent.
■ LOCAL SITES Lower trailsides to Singing Pass, Wedgemount Lake and Rainbow Lake. Flowers mid- to late June; red to dark berries found late August to October.

QUEEN'S CUP or BEAD LILY
Clintonia uniflora • Lily family: *Liliaceae*

■ **DESCRIPTION** Queen's cup is a herbaceous perennial from 8 to 15 cm in height. Its white, cup-shaped flowers grow to 3 cm across, usually with only one borne at the end of a slender stalk. The cobalt blue fruit is round to pear-shaped and singular. The 2 to 5 bright green leaves are broadly lance-shaped, basal and fleshy. The genus name commemorates DeWitt Clinton, governor of New York State, botanist and developer of the Erie Canal.

■ **HABITAT** Cool moist coniferous forests at low to high elevations.

■ **LOCAL SITES** Large patches along trailside to Brandywine Falls, Nairn Falls, Showh Lakes and at mid levels to Garibaldi Lake. Flowers start mid-June at lower elevations; berries turn blue by mid-August until October.

INDIAN HELLEBORE or CORN LILY
Veratrum viride • Lily family: *Liliaceae*

■ **DESCRIPTION** Indian hellebore is a tall, herbaceous perennial from 1 to 2 m in height. Its strongly ribbed, grass green leaves are ovate to elliptic, 10-30 cm long, with a passing resemblance to corn leaves. The numerous yellowish green flowers hang in drooping panicles from the upper portion of the plant.

CAUTION: this plant is extremely poisonous.

■ **HABITAT** Moist open forests at low to high elevations and wet alpine meadows and swales.

■ **NATIVE USE** Indian hellebore was known for its magical power and highly valued by virtually all coastal people. Although it is poisonous, it was used as a medicine and to ward off evil spirits. It was known as a *skookum* ("strong") medicine.

■ **LOCAL SITES** Alpine meadows, Harmony Bowl at Whistler, Seventh Heaven at Blackcomb. Groves can be seen as you descend into Singing Pass from Oboe Summit, along Ancient Cedars Trail on Cougar Mountain and beside trails to Wedgemount and Rainbow lakes. Flowering starts in June at lower elevations and at the end of July at higher elevations.

TIGER LILY

Lilium columbianum • Lily family: *Liliaceae*

■ **DESCRIPTION** Tiger lily is an elegant herbaceous perennial to 1.5 m tall. Its drooping flowers go from deep yellow to bright orange. A vigorous plant can have 20 or more flowers. Shortly after the flower buds have opened, the tepals curve backward to reveal maroon spots and anthers. The leaves are lance-shaped, usually in a whorl and 5-10 cm long. It is said that he or she who smells a tiger lily will develop freckles.

■ **HABITAT** Diverse range, including open forests, meadows, rock outcrops and the sides of logging roads, at low to subalpine elevations.

■ **NATIVE USE** The bulbs were boiled or steamed and eaten.

■ **LOCAL SITES** Common in moist meadows, trailside to Cheakamus Lake, Singing Pass, lower to mid levels to Garibaldi Lake, Nairn Falls. Flowering starts mid-June.

DEATH CAMAS

Zygadenus venenosus • Lily family: *Liliaceae*

■ **DESCRIPTION** Death camas is a herbaceous perennial, to 50 cm in height from bulb. Its small creamy flowers are 1 cm across and neatly arranged in terminal racemes on stems to 50 cm long. The grass-like leaves are mainly basal, to 30 cm long, and have a deep groove like a keel down the centre. The entire plant is poisonous and when out of flower can be confused with the edible common camas (*Camassia quamash*).
CAUTION: the entire plant is poisonous.

■ **HABITAT** Rocky outcrops and grassy slopes at low elevations.

■ **NATIVE USE** The bulbs were mashed and used as arrow poison.

■ **LOCAL SITES** Rare; patches can be seen between Nairn Falls and One Mile Lake. Full flower mid-June.

FALSE SOLOMON'S SEAL
Smilacina racemosa • Lily family: *Liliaceae*

■ DESCRIPTION False Solomon's seal is a herbaceous perennial to 1 m in height. Its fragrant white flowers are borne terminally in 5-10 cm triangle-shaped racemes. The large ovate leaves (10-15 cm long) are stalked and almost clasp the long arching stems. The red fruit grows in terminal clusters and is round, full and abundant.

■ HABITAT Usually seen in large showy patches in moist shady forests at low to mid elevations.

■ NATIVE USE The berries were eaten occasionally, but are not recommended for consumption.

■ LOCAL SITES Common; Brandywine Falls area, mid levels in Singing Pass, large patches beside the nature trail from Whistler Village to Lost Lake. Flowering begins May to June and berries ripen by mid-August.

STAR-FLOWERED SOLOMON'S SEAL
Smilacina stellata • Lily family: *Liliaceae*

■ **DESCRIPTION** Star-flowered Solomon's seal is a smaller, more refined plant than false Solomon's seal (*S. racemosa*; see page 35). It grows to a height of 60-70 cm and has attractive white star-shaped flowers that grow in open terminal clusters. The broad lance-shaped leaves grow on short stalks, alternate, and are 15 cm long; they are usually folded down the midrib and have somewhat clasping bases. The round immature fruit is green with dark stripes and ripens slowly, turning dark bluish black. The star-shaped flowers and striped fruit distinguish this species from other *Smilacina*.

■ **HABITAT** Moist shaded forests, often in association with devil's club (*Oplopanax horridus*) at low to mid elevations.

■ **NATIVE USE** As with false Solomon's seal, the berries do not have much flavour but were eaten on occasion.

■ **LOCAL SITES** Brandywine Falls area; scattered about the forested zones surrounding the lower lakes and at lower levels in Singing Pass; lots to be seen trailside to Cheakamus and Rainbow lakes.

HOOKER'S FAIRYBELLS
Disporum hookeri • Lily family: *Liliaceae*

■ **DESCRIPTION** Hooker's fairybells is an elegant branching herbaceous perennial to 1 m in height. Its white flowers hang in pairs and sometimes in threes, with the stamens extending beyond the petals: this distinguishes Hooker's fairybells from Smith's fairybells (*D. smithii*), whose stamens do not extend beyond the petals. By August the flowers have been replaced by yellow red berries. The common name commemorates Joseph Hooker (1817-1911), a prominent English botanist.

37

■ **HABITAT** Cool moist forests at low elevations.
■ **NATIVE USE** The berries were considered poisonous.
■ **LOCAL SITES** Huge drifts trailside to Cheakamus Lake, lower trails to Garibaldi Lake, along Nita and Alta lakes. Flowers end of May through June.